16th Horizons in Molecular Biology

16TH HORIZONS IN MOLECULAR BIOLOGY

International PhD Student Symposium and Career Fair for Life Sciences

9th-12th September 2019
Göttingen, Germany

Bibliographical information held by the German National Library
The German National Library has listed this book in the Deutsche Nationalbibliografie (German national bibliography); detailed bibliographic information is available online at http://dnb.d-nb.de.
1st edition - Göttingen: Cuvillier, 2019

Nonnenstieg 8, 37075 Göttingen, Germany
Telephone: +49 (0)551-54724-0
Telefax: +49 (0)551-54724-21
www.cuvillier.de

ISBN 978-3-7369-7074-8
eISBN 978-3-7369-6074-9

This publication is printed on acid-free paper.

Booklet by Horizons Organizing team 2019,
InDesign template provided by Sara Osman, Katharina Seitz,
InDesign Layout: Claus-Peter Adam, MPI-BPC,
Group Pictures: Irene Böttcher-Gajewski, *MedienService*, MPI-BPC

Acknowledgements

It took almost twelve months to get us where we are now. Started with a big and extremely motivated team, we organized ourselves in smaller groups, each responsible for another field of organization. All those up and downs, deadlines and unexpected problems, they kept us busy but also motivated. Horizons would not be Horizons if not many things have to be solved last minute by a creative, innovative and smart team. Horizons is a great success, every year, not because we have a lot of experience in organizing conferences, but because that is what we live for in those twelve months before Horizons.

However, burning desire alone does not make a conference. There are many people involved in this adventure, people, which are essential for the success of our conference. Of course, we have to thank Dr. Steffen Burkhardt and Kerstin Grüniger from the Molecular Biology Coordination Office. Without their experience, network, ideas and motivation, Horizons would have never taken place. Many thanks for your input and problem solving skills.

We also want to thank the Georg-August University of Göttingen and the Max Planck Institute for Biophysical Chemistry for their support and for letting us use their space and resources to set up all the talks, booths, posters and events. We specially thank also the directors of the Max Planck Institutes, because without them and their support, the exchange of knowledge and intense interactions with the speakers and the campus would not be possible. We also benefit from the administration people, thanks for processing our questions and reimbursement issues.

As you will see in our booklet, we have again various supporters from Göttingen and other places in Germany, from single people to global companies. Without their help and resources, Horizons would be neither that large nor that entertaining. We would not have food or breakfast, we would not have these great prizes nor anything for the conference bags, not even this booklet. It is fascinating how many different groups and companies are already willing to support this student-organized event. During the last weeks, even more companies wrote to us and asked whether they can participate. This is a good sign, not only for us but also for the entire region of Göttingen and southern Lower Saxony. It is not without reason that even the federal ministry for education and research, represented by Anja Karliczek, honors us with their patronage. Also, thanks to the local media like the Göttinger Tageblatt but also German journals like the Laborjournal, more and more people learn about our symposium. We feel that Horizons is changing the scientific atmosphere in Göttingen and beyond and we hope that this process will progress in the future.

Further thanks to the people "behind the scenes", i.e. Michael Hartig who is the master of sound and light, together with Peter Lösel and the porters. Without you, we would sit in a dark lecture hall and the speakers would have to shout to get heard. We also thank Frederick Köpper, Claus-Peter Adam, Hartmut Sebesse and Irene Böttcher-Gajewski for their professional contribution to our booklet, for making stunning photos of Horizons for years already and for printing our posters, schedules and putting them on the screens. Also thanks to the university and UMG for letting us use their screens to promote this conference. Thanks to Uwe Krüger for serving us delicious food and a wide variety of drinks throughout the conference. We appreciate the work and effort of our MPI drivers. Without you, our honorable speakers would never reach the campus.

Last but not least, we want to appreciate our mentors, previous organizers and friends who are greatly involved in Horizons. Thank you for staying with us all the time and doing work that no one will appreciate as much as we do. Thanks to our colleagues and labs, the material and resources that we were allowed to use and that we can be on this great adventure. Finally, thanks to this enthusiastic team of organizers. It was a wonderful time, we will remember every up and down and we are looking forward toward the next Horizons already, with new energy, with a burning fire and our minds wandering out beyond the Horizon.

From the Horizons Organizing Committee 2019

Table of Content

An International Scientific Conference from PhD students for PhD students

Welcome to the 16th Horizons in Molecular Biology!

Horizons in Molecular Biology symposium was conceived by the graduate students of the *International Max Planck Research School in Molecular Biology* to widen their own horizons beyond the rut of classes and lab work. Ever since, Horizons has grown from a humble idea to one of the most sought-after events in Göttingen and Germany. Horizons aims to bridge the gap between young scientists and experienced researchers by promoting them to engage in a productive dialog and exchange information. This year, we have planned four exhilarating days of presentations, poster sessions, panel discussions, and workshops aimed to inform, inspire and pique the interest of the young, scientific minds. One could describe Horizons as a platform built to facilitate interactions between participants and speakers! We therefore invite you to engage with our roster of distinguished speakers from diverse fields of work during their presentation or even over a cup of coffee. Do not miss this opportunity to network with your favourite speakers and participants from around the world. Should you require an introduction, we are glad to help you! From academia to far and beyond! Our *Career Fair* is specially tailored to meet the needs and expectations of a budding researcher. In addition to our informative talks and educational workshops, we also have a 'Speed Dating' event where you can gain valuable insights from the diverse speakers in a personalized and informal setting. Make sure to sign up for this unique experience!

Life in science can be gratifying and fun, but is often punctuated with times of hardships, doubt and disheartening lows, too. Graduate life serves as a training ground where failures are also important in moulding one as a capable scientist. However an unlucky streak of failures can leave one doubting their caliber. This is when science may become less of a passionate pursuit and more of a job that needs to be done. Being all too familiar with the experience, we have organized a *Panel Discussion on "Being a Scientist: Job or Passion?"*, to get advice from our diverse speakers on how to face such adversities. We have planned the discussion in a highly interactive format so our participants get to ask many questions, too. Learning from the experiences of accomplished scientists is of vital importance, however, here at Horizons, we believe it is equally imperative to support and network with our peers. The Poster Sessions and Awarded Student Talks provide just such an opportunity. We encourage everyone to actively participate in the sessions and vote for whichever poster appealed to you the most! We look forward to having you here with us and hope to make this not only an educational experience, but also a memorable one.

Horizons in Molecular Biology 2019 Organizing Team

Organizing Committee

The 16th PhD Student Symposium *Horizons in Molecular Biology* is organized by a group of PhD students of the *International Max Planck Research School for Molecular Biology* in Göttingen.

Julio Abril Garrido
Max Planck Institute for Biophysical Chemistry
Department of Molecular Biology

Gerrit Altmeppen
Max Planck Institute for Biophysical Chemistry
Department of Meiosis

Polina Derevyanko
University of Göttingen,
Cancer Biology

Gaurika Garg
Max Planck Institute for Biophysical Chemistry
Department of Molecular Biology

Antony Gruness
University Medicine Göttingen,
Department of Pediatrics and
Adolescent Medicine

Katarina Harasimov
Max Planck Institute for Biophysical Chemistry
Department of Meiosis

Ida Jentoft
Max Planck Institute for Biophysical Chemistry
Department of Meiosis

Sakshi Jain
Max Planck Institute for Biophysical Chemistry
Department of Physical Biochemistry

Yi-Tse Liu
University of Göttingen,
Department of Plant Biochemistry

Yen-Yun Lu
University of Göttingen,
Department of Molecular Genetics

Kseniia Lysakovskaia
Max Planck Institute for Biophysical Chemistry
Department of Molecular Biology

Wiebke Maurer
University Medicine Göttingen
Clinic for Cardiology and Pneumology

Vitalii Mudryi
Max Planck Institute for Biophysical Chemistry
Department of Physical Biochemistry

Tarana Nigam
European Neuroscience Institute
Neural Circuits and Cognition Group

Dilantha Perera
Max Planck Institute for Biophysical Chemistry
Department of Cellular Logistics

Anuruti Swarnkar
Max Planck Institute for Biophysical Chemistry
Department of Membrane Protein Biochemistry

Panagiotis Poulis
Max Planck Institute for Biophysical Chemistry
Department of Physical Biochemistry

Taras Velychko
Max Planck Institute for Biophysical Chemistry
Department of Molecular Biology

Sofiia Reshetniak
University Medical Center Göttingen
Institute of Neuro- and Sensory Physiology

Gabriel Villamil
Max Planck Institute for Biophysical Chemistry
Department of Molecular Biology

Debojit Saha
Max Planck Institute for Biophysical Chemistry
Department of Meiosis

Aikaterini Vrentzou
Max Planck Institute for Biophysical Chemistry
Department of Membrane Protein Biochemistry

Ninadini Sharma
Max Planck Institute for Biophysical Chemistry
Department of Meiosis

Roya Yousefi
University Medicine Göttingen
Department of Cellular Biochemistry

Partners

The 16th International PhD Student symposium is financially supported by the following institutions through the International Max Planck Research School for Molecular Biology.

GEORG-AUGUST-UNIVERSITÄT GÖTTINGEN

MAX-PLANCK-GESELLSCHAFT

Donors

The PhD Student Organizing Committee would like to thank the following companies and organizations for their support:

biomol

Cuvillier Verlag

eurofins Genomics

NEW ENGLAND BioLabs® GmbH

GILSON®

iba Solutions For Life Sciences

Program Overview

Monday, September 9th

09:15 *Registration Career Fair*
09:45 **Opening Ceremony Career Fair**
10:00 **Workshop Deb Koen** – *Presenting that Pop: Bringing Science to Life*
10:00 **Simone Mayer** – *Hertie Institute for Clinical Brain Research*
10:30 **Michael Becker** – *Bayer*
Coffee Break
11:20 **Christine Mieck** – *Springer Nature Communications*
11:50 **Andreas Laustsen** – *Technical University of Denmark & Biotech Entrepreneur*
12:20 **Speed Dating /** *Lunch break*
13:00 **Workshop Angelika Hofmann** – *The Art of Grant Writing*
13:30 **Beata Mierzwa** – *Beata Science Art*
14:00 **Magdalena Wienken** – *Definiens GmbH*
14:30 **Bernd Reichert** – *EU Commission*
Coffee Break
15:30 **Horizons Opening Ceremony**
15:45 **Michael Rosbash** – The Circadian Rhythm Story: Past, Present and Future
16:45 **Robert Ernst** – Emerging roles of the Unfolded Protein Response in membrane homeostasis
17:30 **Argyris Papantonis** – Impaired chromatin remodelling links misregulation of autophagy and DNA damage response to a rare Progeroid Syndrome

19:30 *Join us for a beer!* (Z.A.K., Am Wochenmarkt 22, 37073 Göttingen)

Tuesday, September 10th

09:00 *Registration*
09:30 **Randy Hampton** – Mallosteric Misfolding and Rhomboidal Retrotranslocation: Lessons From Regulated ERAD
10:15 **Gaia Pigino** – Mechanistic understanding of cellular processes by cryo-EM
Coffee Break
11:30 **Christopher Vollmers** – Sequencing for transcriptome analysis
12:15 **Leonie Ringrose** – Epigenetics meets mathematics: The fusion of experiment and theory brings insights beyond intuition
Lunch
13:50 **Miki Ebisuya** – Human Time vs. Mouse Time: *In vitro* segmentation clock as a model system
14:35 **Needhi Bhalla** – A tale of two checkpoints
Coffee Break
15:35 **Awarded Student Talk**
15:55 **Zoya Ignatova** – Translational control: probing dimensionality beyond linear sequence of mRNA
16:40 **Asifa Akhtar** – Epigenetic regulation by histone acetylation
17:25 **Poster Session 1** ***(wine and cheese)***

Wednesday, September 11th

09:00 *Registration*
09:30 **Jen Heemstra** – A chemical biology toolbox for specific RNA modification and capture
10:15 **Tatjana Kleele** – Degradation and proliferation are outcomes of distinct types of mitochondrial division
Coffee Break
11:30 **Sjors Scheres** – The atomic structures of Tau filaments from Alzheimer's and Pick's disease brains
12:15 **Max Cryle** – Understanding the biosynthesis of the glycopeptide antibiotics
13:00 **Poster Session 2 / *Lunch break***
15:00 **Margaret McCarthy** – Surprising origins of sex differences in the brain
15:45 **Pavel Tomancak** – A new force awakens: comparative approach to tissue morphogenesis in insects
Coffee break
16:50 **Awarded Student Talk**
17:10 **Leo James** – Epigenetic regulation by histone acetylation

19:15 City tour

20:15 Horizons Dinner (*Paulaner,* Düstere Straße 20A, 37073 Göttingen)

Thursday, September 12th

09:00 *Registration*
09:30 **Food for Thought:** *Horizons Breakfast*
10:15 **Naama Barkai** – Gene transcription as a limiting factor in protein production and cell growth
Coffee Break
11:00 **Awarded Student Talk**
Break
11:40 **Panel Discussion: *Being a scientist: Job or Passion?***
Lunch & Coffee
14:30 **James Williamson** – Ribosome Assembly in Bacteria
15:15 **Michael Levine** – Enhancer-promoter communication in living *drosophila* embryos

16:00 **Closing Ceremony**
16:25 **Champagne and Goodbye**

GENERAL INFORMATION

Admission to Plenary Lectures

Admission to plenary lectures is restricted to registered participants.

Name Badges and Conference Package Vouchers

You will receive your name badge and vouchers for the conference package (if you chose it) during registration. The conference package includes lunches, city tour and conference dinner and party. Please wear your name badge at all times during the symposium. It will be required for admission to both scientific and social programs.

Poster Sessions

If you are presenting a poster, you will be given your poster number during registration. Make sure that you put up your poster on the right poster stand. The number will serve as an identification mark for our poster committee. Pins for putting up the poster will be available at the registration desk.

Poster Session I: Tuesday, 17:25-18:55, *(Wine and Cheese),*
Poster Session II: Wednesday, 13:00-14:30, *(Lunch Buffet).*

Poster Prizes

Prizes will be awarded to the top two participants selected by our juries. All registered participants presenting a poster are eligible for the prize. All participants must be available during the first poster session and present their poster to a jury of peers. Short-listed posters will be marked with a ribbon the following morning and these participants are expected to present their work to a jury comprised of Speakers and IMPRS faculty members.

Additionally, a prize will be awarded to the participant whose poster receives the most votes in our anonymous 'People's Choice Award' ballot. So be sure to drop off your ballot (found in your nametag) in one of our ballot boxes!

Travel Grants

One of the main objectives of Horizons is to give young scientists a platform to present and share their research findings at various stages of their career. The awarded student talks give selected PhD students the opportunity to present their work in the plenary lecture of the conference. This year we received more than 50 applications from all over the world representing 21 countries over 5 continents. After a rigorous selection process, we would like to congratulate this year's awardees.

Awarded Student Talks

- **Angel Santiago-Lopez (10th September, 15:35)**
  Mobilization of cellular stress signaling in protein misfolding diseases for controlled gene delivery - *Georgia Institute of Technology, USA*

- **Julie Trolle (11th September, 16:50)**
  Engineering Prototrophy In Mammalian Cells - *New York University, USA*

- **Triana Amen (12th September, 11:00)**
  Stress Granules regulate cell metabolism through clustering of mitochondrial membrane protein VDAC - *Universitätsmedizin Göttingen, Germany*

Poster Presentation

- **Tomas Lagunas** - *Washington University School of Medicine*, USA
- **Rima Siauciunaite** - *Karlsruhe Institute of Technology, Germany*
- **Priscilla Gonzalez** - *Pontificia Universidad Catolica de Valparaiso, Chile*
- **Ivana Borovska** - *Slovak Academy of Sciences, Slovakia*
- **Abhishek Upadhyay** - *Humboldt University of Berlin, Germany*
- **Alina Stein** - *Universität Basel, Switzerland*

Diversity of applications received for Horizons 2019

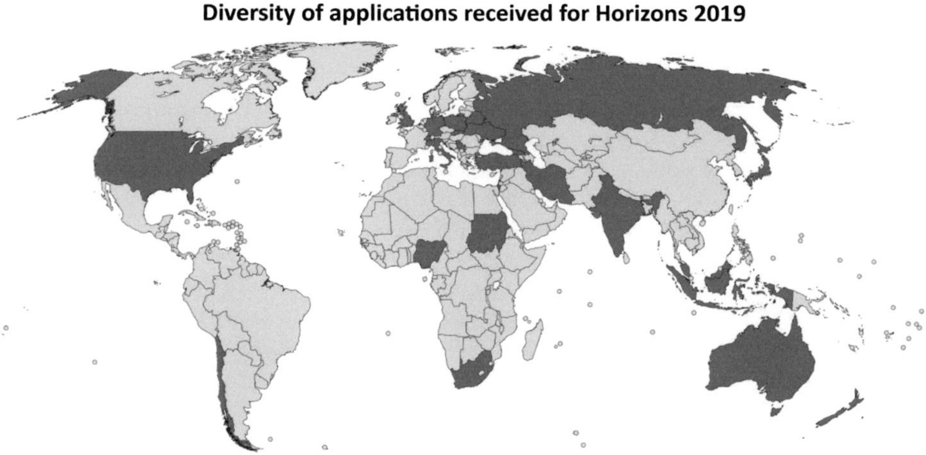

Social Events

Horizons is not only about scientific exchange but also about building personal contacts with people in your field. A long standing tradition of the symposium is to organize social events so that the participants and speakers get a chance to interact in an informal setting. This year we will be hosting the following activities:

Social Mixer: Join us for a Beer!

After an engaging day of career fair and scientific discussions, we invite the participants and speakers to join us in exploring the colorful nightlife of Göttingen. We will visit one of the local's favorite bars to experience the lively atmosphere. It is a great opportunity to meet new people over a glass of traditional German beer and food from 19:30 on Monday, September 9th . All participants are invited to join, the event is not included in the conference package.

Location: *Z.A.K.* (Am Wochenmarkt 22, 37073 Göttingen)

Wine and Cheese

During one of the poster sessions, a wine and cheese tasting is organized, where a great variety of cheese from different places will be served. With a glass of wine in hand, you can have forthright discussions about the research of young scientists from all parts of the world. Tuesday, September 10th, 17:25 (during Poster session 1)

City Tour

Göttingen is known as the city of Science, it dates back to the Middle Ages and boasts its well preserved half-timbered houses. Join us in a short city tour combining the fantastic architecture and the scientific historical overview. The guided tour will take place on Wednesday, September 11th, starting at 19:15 h. The tour is in English. We will meet at the Gänseliesel fountain in the city center and end the tour at *Paulaner* where we will have the conference **dinner and party**. The tour is for participants who opted for the conference package and the speakers.

Starting point: Gänseliesel fountain in the city center

Dinner and Party

As every year, we invite you to a night of informal interaction with dining and merriment. It is a good occasion to get to know the other participants and speakers and to celebrate new friendships over a couple of drinks, food, games and music and dancing starting at 20:15 on the 11th of September (only included in conference package).

Location: *Paulaner - Zur alten Brauerei Göttingen* (Düstere Straße 20A, 37073 Göttingen)

"Food for Thought" Breakfast

On Thursday morning we invite you to our science breakfast, to enjoy bread, pastries and Castagnes with a cup of coffee or tea on our last conference day! We thank our sponsors *Ruch* and *Küsters* for sponsoring the breakfast.

History of Horizons

December 4th, 2003. The seminar room in the Göttingen Center for Molecular Biosciences was bustling with activity. Organizers darted in and out of the room, making sure everything was in order and as it should be. Speakers attempted to look through their talking points, whilst entertaining the many curiosities of the next generation gathered around them. The graduate students themselves, whispered to each other and waited apprehensively, with thinly veiled excitement, for the program to begin. There were just under a hundred eager minds crowded into that one room. Students and researchers; all brought together by a shared passion: a passion for science. This was the inaugural **Horizons in Molecular Biology Conference**: An International PhD Student Symposium!

The conference has come a long way since then. Over the years, the number of participants grew steadily and increased in variety. Horizons now regularly attracts around *three hundred participants from over thirty countries*. The symposium has featured numerous renowned researchers, representing a wide range of fields in the life sciences. This exhaustive list also included several Nobel Laureates, namely: Professors **Martin Chalfie**, **Sir John Walker**, **Ada Yonath**, **Thomas Südhof**, **Kurt Wüthrich**, **Carol Greider**, **Venki Ramakrishnan**, **Sir John Gurdon**, **Tim Hunt**, and this year **Michael Rosbash**.

Horizons has also evolved continuously as each new batch of student organizers brings a fresh perspective and new ideas, proving consistently that you *can* make a good thing better! In 2006, the organizers introduced a **Poster Session** accompanied by **Student Talks**. PhD students presented *over a hundred posters* and a select few were given the chance to present their work to an audience of peers and professors. Building on this, in 2007, Horizons launched its first **Career Fair** in conjunction with the conference. The fair offered budding scientists with an opportunity to connect with industry and network with life science professionals with a different backgrounds and perspectives. Representatives from over *twenty companies* were present at the inaugural Career Fair, conducting live interviews and CV checks. Today, the Career Fair has adopted a major role in the conference, offering **workshops** held by professionals, and interactive discussions with key figures in the modern life science industry. All catered to the needs of a young scientist.

At the heart of Horizons, is an idea: **An International Scientific Conference *from* PhD students *for* PhD students**. Organized by students from Göttingen's Intetrnational Max Plank Research School for Molecular Biology, Horizons in Molecular Biology provides an interactive experience in a relaxed environment. It aims to bridge the gap between young scientists and experienced researchers and bring together experts and novices from various fields of the life sciences to engage in a productive dialog and exchange information. Modern discovery is persistently progressing from exploring rainforests in faraway lands to observing the nature of life under a microscope. We are the modern explorers, and are committed to keep progressing forward, in the pursuit of science, towards the Horizon.

IMPRS for Molecular Biology

For the 16th time, the Horizons symposium is organized by students of the International Max Planck Research School (IMPRS) for Molecular Biology at the University of Göttingen. The Molecular Biology Program is jointly conducted by various departments of the University of Göttingen, the University Medical Center Göttingen, the Max Planck Institutes for Biophysical Chemistry and Experimental Medicine, and the German Primate Center. The common aim is to offer an intensive, research-oriented education in order to prepare the students for a professional career in the life sciences. This broad spectrum of topics is mirrored in the fields covered by Horizons. In addition to the small classes and specialized hands-on laboratory courses, the intercultural experience is extraordinary. Excursions, culture nights and workshops with students from all over the world naturally become a "seminar" on intercultural communication. Students can join the program after completing their Bachelor studies. During the first year, students receive a broad education in molecular biology through lectures, lab rotations and methods courses delivered by the participating institutions. From the second year, the students conduct their own research during their MSc or PhD thesis projects and receive further training in specialized methods and skills courses within the Göttingen Graduate Center for Neurosciences, Biophysics and Molecular Biosciences (GGNB), which is funded by the German Excellence Initiative.

In September 2000 the first brave students from as far away as Ghana, China, Mexico and Malaysia came to Göttingen to join the new International MSc/PhD Molecular Biology Program. Today, with the twentieth batch of twenty students joining the IMPRS, the program has become very successful as demonstrated by excellent evaluations, awarded prizes, and applications from all over the world. By now, more than 175 students have successfully defended their PhD theses within the program and kissed the Gänseliesel. Kissing this statue that stands on a fountain on the market place in the middle of the city and giving her flowers is an old tradition for Göttingen PhD students. For more detailed information about the International MSc/PhD Molecular Biology Program in Göttingen, please take a look at our website:

www.gpmolbio.uni-goettingen.de

About Göttingen

The renowned university town of Göttingen lies in the center of Germany in southern Niedersachsen (Lower Saxony) between the Harz mountains and the Weser river. Its establishment dates back to 953, it was chartered in 1210 and flourished as a member of the Hanseatic League.

A glimpse of this hanseatic opulence is still present today in the medieval Town Hall, the splendid churches and quaint half-timbered houses in Gothic, Renaissance and Baroque styles. Göttingen's landmark is the "Gänseliesel", an art nouveau statue of a girl herding geese on top of the fountain on the market square, outside the Old Town Hall. Traditionally, all new doctoral graduates of the University kiss the cheeks of the statue after passing their examinations, making her "the most kissed girl in the world".

The Georg-August University was founded in 1734 and today has 13 faculties with 32,000 students. The university sends forth many famous scientists – amongst its alumni are 45 Nobel laureates. Especially many famous mathematicians come from Göttingen: Carl Friedrich Gauss, Bernhard Riemann and David Hilbert, to name only a few.

Apart from the university, the city is also home to many research institutes, such as Max Planck Institutes, and international companies.

Nearly untouched by bombings in World War II, the city center is now an attractive place to live in with many shops, small cafes and chic bars. Consequently many students live here, giving Göttingen a young face: in 2003, almost half of the population of the inner city was between 18 and 30 years of age. Today Göttingen has approximately 130,000 inhabitants including 24,000 students, giving it the charm of a small and lively university city.

Practical Information

Certificates of attendance will be handed out during registration upon request.

Insurance

The Organizing Committee accepts no responsibility for accidents or damage to participants' private property. Please make your own arrangements for all the necessary insurance.

Internet

Free Wi-Fi internet (eduroam) is available throughout the symposium in the lecture facilities.If you do not have access to eduroam, please contact the registration desk.

Parking

Participants can use the parking lots available near the MPI for Biophysical Chemistry.

Buses

The buses number 21, 22 and 23 to the city center leaves just outside the symposium venue at the bus stop „Faßberg". Bus schedules are available at the registration desk.

Taxi

Taxis to the city center can be arranged by calling one of the many Göttingen taxi companies:

- *Göttinger-Funk-Taxi-Zentrale:* 0049-551-69300
- puk minicar: 0049-551-484848

Please note that taxis are considerably more expensive than public transport. Also, keep in mind that taxis only accept cash.

Organizers

In case you have difficulties ordering a cab, getting around town or else, you are always welcome to ask for help one of the organizers. Just look out for students wearing a **Horizons t-shirt** and get in touch with us regarding your issue.

Services

Registered participants are entitled to the following: Admission to plenary lectures, conference material (certificate of participation, name badge, conference bag) and refreshment during coffee breaks.

Participants who opted for the conference package, in addition to the above get a voucher for lunch, city tour, conference dinner and party and a hardcopy conference booklet.

Career Fair - Schedule

16TH HORIZONS IN MOLECULAR BIOLOGY

Max Planck Institute for Biophysical Chemistry, Göttingen, Germany
9-12 September 2019

MONDAY, 9th

9:15 Registration Career Fair

9:45 Opening Ceremony Career Fair

10:00 Simone Mayer
Hertie institute for Brain Research

10:30 Michael Becker
Bayer

11:00 Coffee Break

11:20 Christine Mieck
Nature Communications

11:50 Andreas Laustsen
Technical University of Denmark & Biotech Entrepreneur

12:20 Speed Dating
Lunch

13:30 Beata Mierzwa
Beata Science Art

14:00 Magdalena Wienken
Definiens GmbH

14:30 Bernd Reichert
EU Commission

15:00 Coffee Break

workshops

10:00 Deb Koen
Presentations that Pop: Bringing Science to Life
→ 12:30

13:00 Angelika Hofmann
The art of grant writing
→ 15:30

Welcome to the 13th Career Fair for Life Sciences

Your PhD is the starting point of your career in science. But have you ever found yourself asking: what is a career in science? A Post-Doc and tenure track to become a group leader? A position in a big biotech company? What are my options?

The answer is that with a PhD within the natural sciences, you have an ocean of options. Certainly more than two! The truth is, during a PhD you will acquire skills and a mind-set sought after everywhere.

In this year's Horizons Career Fair, we introduce you to seven people, all with a background in the natural sciences, who have chosen seven different career paths. These speakers represent the fields of science policy, academic research, industry, startups and entrepreneurship, publishing, consulting, and science art. Our goal is to present a broad range of possible future paths for our participants and to create an informal environment in which students can explore their possibilities after a PhD or a Post-doc in the natural sciences.

In addition, we give you the opportunity to learn the "dos" and "don'ts" of research proposal writing and a chance to learn clever tricks to become a more confident speaker in our two workshops.

The Horizons career fair 2019 will give you a taster of what opportunities are out there and broaden your Horizons!

Career Fair Organizers

Notes

Monday 10:00 am

Insights: Academia

Following one's own curiosity and pushing forward the frontiers of human knowledge is a dream for many PhD students. But how do you make this dream come true? And is the grass really so much greener on the other side? Simone Mayer will share her strategies and decision points that crucially contributed to her long-term goal of becoming a principal investigator. Having only recently started her own research group, she will share the unexpected joys as well as the unforeseen hurdles involved in this endeavor. For Simone, heading your own research lab is a bit like having your own company. As the CEO, one does a diverse array of jobs including being the head of Human Resources, the Head of Acquisitions, the Head of Public Relations as well as the Head of Finance and Research Strategy. For these reasons, heading her own group has been a very fun experience that has allowed Simone Mayer to grow as a person in many different areas already. Simone Mayer is looking forward to developing her research endeavor further in the coming years to push the frontiers of human knowledge on human brain development and evolution.

Simone Mayer

Hertie Institute for Clinical Brain Research

Dr. Simone Mayer is a molecular neurobiologist interested in understanding the development of the neocortex, the brain structure that has significantly contributed to the evolution of advanced cognitive functions in mammals. Simone conducted her doctoral work at the Max Planck Institute of Experimental Medicine focused on inhibitory synapses. Being interested in both academia and science policy, Simone decided to get practical experience beyond the lab during a traineeship at the European Research Council Executive Agency in Brussels, Belgium. After 5 months in an office, however, she was happy to be back at the bench starting her postdoctoral work at the University of California, San Francisco. Being based in the Bay Area, in collaboration with a biotech company, Simone established a novel method to determine physiological and transcriptomic features of single cells simultaneously. Since September 2018, Simone is heading an independent research group at the Hertie Institute for Clinical Brain Research in Tübingen, Germany. Her research questions focus on how environmental impacts affect brain development, for example by modulating the neurotransmitter system.

Michael Becker

Bayer

Michael Becker is heading a laboratory in the area of kidney disease research at Bayer AG. In addition, he is leading multiple preclinical research projects partly involving other companies or academic institutions. After studies in Biochemistry at Tübingen University and Harvard Medical School, he did his PhD at the German Cancer Research Center in Heidelberg. He joined Bayer in 2015 in the area of Lead Discovery, where he worked on uHTS assay development and switched to his current position in 2016. In addition to his passion for drug discovery he has a strong interest in innovation. He won the Merck Innovation Cup in 2014 and is supporting innovation at Bayer in his role as Innovation Coach.

Monday
10:30 am

Preclinical Research in the Pharmaceutical Industry

Michael will talk about his personal career path and decisions who led him to his current position. Moreover, he will give an insight into his everyday work and the differences between research in pharma and academia. At the end of his talk, perspectives into other positions in pharma R&D and beyond will be given.

Monday 11:20 am

Insights: Publishing

Do you love science but feel that a career at the bench isn't enough to sate your desire to learn more about the natural world? Do you enjoy reading papers outside your chosen area of research? If the answer is 'yes' to any of these questions, consider an editorial career!

To give you an insight in the publishing world and help you decide whether this could be a career choice, Christine will talk about recommended skills for being an Editor and how the daily life at Nature Journals looks like. Please bring all your questions regarding career options with publishers, editorial workflows and decision making along.

Christine Mieck

Nature Communications

Christine obtained her PhD from the IMP Vienna where she studied kinesin motors using in vitro reconstitution and structural biology approaches. Fascinated by 'bottom-up' approaches to rebuilt dynamic biological systems, she decided to move from motor proteins to studying the dynamic organization of membrane proteins. Her post-doctoral work at the IST Austria explored the biochemical and biophysical bases of cellular polarity in an in-vitro setup. After starting her editorial career as an Editor at Scientific Reports for 3 months, she joined Nature Communications in September 2017. Christine handles submissions on the biochemistry and structural biology of membrane proteins, computational structural biology, and general and single molecule biophysics. Christine is based in the Berlin office of Springer Nature.

Andreas Laustsen
Technical University of Denmark & Biotech Entrepreneur

Associate Professor Andreas H. Laustsen heads the Tropical Pharmacology Lab at the Department of Biotechnology and Biomedicine, Technical University of Denmark. The Tropical Pharmacology Lab is an academic research group focusing on developing biotherapeutics and diagnostics against neglected tropical diseases. Andreas himself is specialized in antibody discovery, toxinology, antivenom, and neglected tropical diseases, and holds a PhD from the University of Copenhagen. He is an Advisor to the World Health Organization's Working Group on Snakebite Envenoming, and he is a co-founder of the biotech companies Biosyntia, VenomAb, Antag Therapeutics, Chromologics, Bactolife, and VenomAid Diagnostics. Andreas is recognized as Denmark's Coolest Engineer, and he was on Forbes 30 under 30 list for 2017 and MIT Technology Review's list of the 35 Top Innovators under 35 in Europe 2017. Andreas spends most of his time with academic engagements, but also still plays an active role in most of the companies he co-founded.

Monday
11:50 am

Venturing into biotech entrepreneurship as a young academic

Upon graduation, the majority of PhD graduates nowadays transition from a daily work life in academia to one in industry. This switch may involve a steep learning curve. However, often PhD graduates will have easy access to mentors and older colleagues that may help them make this transition. In contrast, venturing into biotech entrepreneurship is completely unchartered territory for many young graduates, who may not as easily find access to advice and mentorship. This is both because a much smaller number of biotech entrepreneurs exists in comparison with biotech industry professionals, but also because the young biotech entrepreneur will typically have to work in a very small startup team with only few co-founders and colleagues due to the often limited resources of a young biotech startup. In my talk, I will present my own story from how I co-founded my first biotech company (Biosyntia) during my master studies in 2012, over why I choose to go back to university to pursue a PhD, and end with describing why and how I chose to pursue a career as a university academic, while co-founding 5 additional companies alongside my academic career. I will reflect on the benefits and drawbacks of venturing into (biotech) entrepreneurship, as well as continuing your career in academia, and present my views on how one can merge or bridge these fields. Finally, I hope to share some personal advice for aspiring biotech entrepreneurs, as well as younger academics..

Monday
1:30 pm

Beata Mierzwa

Beata Science Art

Communicating Scientific Concepts Through Art

Creativity is an integral part of both science and art, and using it to combine these two seemingly different disciplines creates unique and effective ways to communicate scientific information. Along my classical academic career path, I have been exploring the visual aspect of science communication using hand- drawn illustrations and science fashion. My science art aims to convey complex biological concepts using both abstract imagery and real scientific data. Each illustration requires breaking down the essence of complex scientific findings and translating them into aesthetic visuals using metaphors that describe abstract biological processes in intuitive ways. This form of visualization provides a powerful tool to facilitate communication between scientists from diverse fields, spark fascination and curiosity, as well as inspire future scientists. In this talk I will share my journey towards combining science and art, and discuss how cultivating passions outside of the lab creates beautiful and unexpected opportunities with the potential to open up new and creative career paths.

Beata is a postdoctoral researcher in the labs of Karen Oegema and Arshad Desai at the Ludwig Institute for Cancer Research and University of California San Diego, where she explores the variation of the cell division machinery in different cell types. After studying Molecular Biology at the University of Vienna, Beata completed her master's thesis at the ETH Zürich. During her PhD in Daniel Gerlich's lab at the Institute of Molecular Biotechnology (IMBA) at the Vienna BioCenter, she focused on the final step in cell division, exploring the dynamics of the ESCRT-III machinery that separates the cells during cytokinetic abscission.

Beata made her first science art drawing to depict her own PhD research, and her illustrations quickly developed into a mission to communicate science. She started spending her evenings drawing in addition to her daily experiments in the lab, making artwork for many conferences, journal covers and research groups. She also designs science-inspired clothes printed with drawings and microscopy images. You can find a gallery of her work at www.beatascienceart.com or follow her on social media *(@beatascienceart)*.

Magdalena Wienken
Definiens GmbH

Monday
2:00 pm

Disrupting Biomedical Science Through Data and AI

Magdalena received a B. Sc. and Honours degree in Applied Biology with a focus on Genetics and Immunology from the University of Applied Sciences in Bonn as well as the University of Aberdeen in Scotland. She continued her studies with an M. Sc. in Molecular Medicine at the University Medical Center Göttingen and finished her PhD in Molecular Oncology as a scholar of the GGNB graduate school. Within her PhD she focused on the p53 independent functions of Mdm2 as a stemness promoting factor and chromatin modifier. After her PhD, Magdalena decided to pursue her career in strategic consulting, working for Accenture Technology Strategy with a Pharma and Biotech focus. She specialized in data-driven strategies, working for the top-ranked pharma companies in the DACH region. She only recently left consulting, continuing her career as a manager at Definiens , a middle-sized, Munich based company, which specializes in tissue phenomics in the context of innovative immunotherapy of cancer.

Finishing a PhD in Science raises a lot of questions and opens many different doors. Am I going to stay in academia? Shall I change into industry? Which alternative job opportunities are out there? Magdalena decided to leave academia for consulting due to 2 reasons: 1) get a broader understanding on the current challenges and opportunities of the Pharma industry and 2) get the opportunity to quickly gain industry and strategic expertise, and work with top management clients in innovative and short-cycled projects.

This talk will outline the pros and cons of a career in consulting, experiences as well as the challenges that might come along with it. In addition, it will highlight why she decided to change back to science again.

Monday 2:30 pm

Bernd Reichert

EU Commission

Eurocrat in spite of oneself, or; love yourself and it doesn't matter who you marry

A short outline of my path into the European institutions and why I stayed there for over 25 years. In this talk, Bernd Reichert will talk about the work within the European Union and the prospects of young scientists within the field of science policy.

University studies and PhD in physical chemistry and quantum physics at the Humboldt University of Berlin. After Post-Doc studies at the Technical Universities of Berlin and Munich, he joined the European Commission, Directorate General for Research, in 1992. There he has held several assignments as programme officer in the areas chemistry, biotechnology, nanotechnology as well as science and society, until he became Head of Administration of the directorate "Industrial Technologies" in 2003. From 2007 until the end of 2013 he was heading the unit "Small and Medium-Sized Enterprises" in the Commission's Research and Innovation Directorate General. Until the end of 2018 he was responsible for setting-up the activities "Horizon 2020: Small and Medium-Sized Companies", in particular the implementation of the SME Instrument under the EU Research and Innovation Framework Programme Horizon 2020, within the Executive Agency for Small and Medium-Sized Companies (EASME).

Deb Koen

Account manager and career coach for Corning Incorporated, and Senior consultant for HR Works, Inc. and for BadFish Consulting.

Deb Koen, Career Strategist and Professional Development Coach, is a recognized leader in the field of career development. She conducts peak performance programs to develop individual and organizational potential. Along with having been a preferred presenter for The Wall Street Journal's Executive Diversity Career Fairs, Deb coaches, consults and presents on career management, transitions, job search strategies, and women in the workforce. She regularly conducts leadership development programs for the International Visitors Leadership Program (IVLP) sponsored by the U.S. Department of State. Deb brings her expertise and enthusiasm to each and every engagement to create innovative solutions.

Deb previously led the private nonprofit organization, Career Development Services, and she has consulted for numerous corporations and counseled in higher education. She is currently the account manager and career coach for Corning Incorporated, and senior consultant for HR Works, Inc. and for BadFish Consulting.

Deb is a co-author of Career Choice, Change & Challenge (Jist, 2000). She has been a columnist for The Wall Street Journal's WSJ.com, for the business section of Gannett's Democrat & Chronicle and for the scholarly scientific journal, Nature. She holds a certificate in nonprofit management from Harvard Business School and M.A. in Counseling from Colgate University. Deb also serves on the Board of Trustees for The Nature Conservancy. In 2016, she received the Corporate Enlightenment Award from Rochester Global Connections for her work with emerging leaders "for exemplifying the ideals of inclusiveness, respect for diversity, global understanding and citizen diplomacy."

Monday 10:00 am

Presentation that Pop: Bringing Science to Life

Communicating science through presentations plays a large role in delivering messages within the scientific community and beyond to the general public. The most effective speakers can design and present a scientific talk that conveys scientific information of value in a way that engages the audience.

This workshop provides practical approaches that are immediately applicable to:

- Prepare the content and delivery for a meaningful and lively scientific presentation that best represents you and your research
- Structure the introduction, core content, and conclusion to engage the audience aurally and visually
- Identify ways to adapt your formal presentation to more casual situations (e.g. networking and informal conversations) to tailor your message and reflect favorably on your career.

Join in this opportunity to increase your confidence and your impact through these key presentation practices.

Monday 1:00 pm

The Art of Grant Writing

For a successful grant, basic grant writing principles are a prerequisite, as is clear scientific writing. Learn to organize and compose your thoughts and apply winning writing principles within your next proposal through this workshop. Participants will also hear about common mistakes to avoid and receive guidance on best practices and the review process. Key sections of a grant will be highlighted, and hands-on exercises will be included. Emphasis will be on how to achieve maximum impact and success in grant writing with the reader and reviewer in mind. The workshop is presented by Dr. Angelika Hofmann, Strategic Projects and Communications Advisor to the Vice Provost for Research at Yale University and author of two major textbooks on the topic: Scientific Writing and Communication – Papers, Proposals, and Presentations (Oxford University Press) and Writing in the Biological Sciences (Oxford University Press.)

Angelika Hofmann

Office of the Vice Provost for Research

As Strategic Projects and Communications Advisor, Angie Hofmann works closely with Yale's Vice Provost of Research in coordinating the implementation of academic priorities in the sciences across the University. In her role, she also leads strategic communications efforts for the research team and is charged with advancing research-related training. In addition, she acts as the liaison on science initiatives between the office of the Vice Provost for Research and various University offices as well as external institutions.

Before joining the VPR team in 2018, she was Director, Corporate and Foundation Relations, Science and International Initiatives, in Yale's Office of Development where she managed some of Yale's most important prospects and played a key role in fundraising from leading international corporations and foundations. As part of this position, she also oversaw a team of professional scientific writers and the operational management of the Corporate and Foundation Relations office.

Angie holds a PhD in molecular biology and biochemistry from the University of California, Irvine, and was a post-doctoral fellow at the Max-Planck-Institute of Molecular Genetics in Berlin, Germany. She is renowned in the scientific writing field and teaches workshops on scientific communication around the world. She is also the author of two books that have become standards in the field: Scientific Writing and Communication – Papers, Proposals, and Presentation (Oxford University Press, 3.ed., 2016) and Writing in the Biological Sciences (Oxford University Press 3.ed., 2018).

Speed Dating

Along with talks and exciting workshops, we have a 'Speed Dating' event in the lineup. Here, the participants get a chance to talk to the speakers of choice face to face. During the lunch break, this event gives you an opportunity to interact with the speakers in a personal and informal environment. In small groups, you will get to meet two speakers and discuss for half an hour all the questions that pop in your heads.

PLENARY LECTURES

Keynote Speaker

Bioinformatics

Cell Biology

Developmental Biology

Molecular Genetics

Imaging

Membrane Biology

Structural Biology

Synthetic Biology

Michael Rosbash

Brandeis University
Waltham, Massachusetts, USA

Monday 3:45 pm

Michael Rosbash is a Professor of Biology and the Peter Gruber Professor of Neuroscience at Brandeis University. He is also an Investigator of the Howard Hughes Medical Institute. Rosbash went to the Newton public schools in greater Boston and then to Caltech, graduating in 1965 with a B.S. in Chemistry. He spent the 1965-1966 academic year in Paris as a Fulbright Scholar in the lab of Marianne Grunberg-Monago and then entered the Ph.D. program at MIT in the fall of 1966. Rosbash worked there in the lab of Sheldon Penman and received a Ph.D. in Biophysics in 1970. After a brief stint at the University of St. Andrews, he was a post-doc in the lab of John Bishop in the Department of Genetics at the University of Edinburgh from 1971-1974. Rosbash joined the faculty of Brandeis University in 1974 and was promoted to Professor of Biology in 1986. He became a Howard Hughes Medical Institute Investigator in 1989.

Rosbash has made fundamental contributions to our understanding of the post-transcriptional regulation of gene expression, especially RNA metabolism in yeast. He is best known however for his work in Drosophila that illuminated our current understanding of the molecular mechanisms that underlie circadian rhythms, the intrinsic clock that controls the cyclic behaviors of all animals. These same molecules, molecular machines and biological principles not only control Drosophila circadian clocks but also the ubiquitous process of circadian rhythmicity throughout

The Circadian Rhythm Story: Past, Present and Future

The last 35 years has seen a sea change in the field of circadian rhythms. This modern era began with work in Drosophila (fruit flies), which has been a leading genetic system for more than 100 years. My colleagues and I discovered the clock mechanism that underlies circadian timing, and it turned out that the genes and mechanism are conserved in all animals. This circadian system governs a large fraction of all gene expression, once again extending from fruit flies to humans, which explains why so much animal physiology (biochemistry, metabolism, endocrinology, behavior, sleep, etc.) is under temporal control. The broad reach of circadian biology indicates that it will continue to be important to many aspects of human well-being and will become increasingly relevant to medicine as more knowledge and applications accrue. Time permitting, I will also touch on the role of the Drosophila brain clock in controlling sleep and locomotor behavior, which remains a central focus of my current lab.

the animal kingdom. This circadian clock also controls much of cell physiology and metabolism, again in all animals - from humans to Drosophila (fruit flies).

Rosbash and his Brandeis colleague Jeff Hall as well as Mike Young of the Rockefeller University have received numerous awards for their circadian work, including most recently the 2017 Nobel Prize in Physiology or Medicine. They previously received the Shaw Prize in Life Science and Medicine (2013), the Wiley Prize in Biomedical Sciences (2013), the Massry Prize (2012), the Canada Gairdner International Award (2012), the Louisa Gross Horwitz Prize for Outstanding Basic Research (2011), and the Peter and Patricia Gruber Foundation Neuroscience Prize (2009). Rosbash also received the Caltech Distinguished Alumni Award (2001), and he is a Member of the National Academy of Sciences, a Fellow of the American Association for the Advancement of Sciences and a Fellow of the American Academy of Arts and Sciences.

Robert Ernst

Saarland University
Saarbrücken, Germany

Monday
4:45 pm

Robert Ernst studied Human Biology at the Philipps-University Marburg and moved for his Diploma thesis to the Goethe University Frankfurt, where started his PhD thesis at the Institute of Biochemistry. He defended his thesis at the Heinrich-Heine University Düsseldorf in 2007. He was a postdoctoral fellow at the Whitehead Institute for Biomedical Research in Cambridge, MA before joining the group of Kai Simons at the Max-Planck Institute for Molecular Cell Biology and Genetics in Dresden, Germany. He then started his own group as Emmy-Noether fellow at the Goethe University Frankfurt. Since 2017, he holds the Chair for Molecular Biology at the Medical Faculty of the Saarland University. His laboratory is interested in how the organelles of eukaryotic cells can sense their own composition and physical properties in order to maintain their identity. Using cross-disciplinary approaches that combine cell biology, biochemistry, biophysics, and theoretical physics, his group uncovered the molecular mechanism of central machineries controlling key aspects of membrane biogenesis: 1) The decision between making more membranes or storing fat, 2) The proportion of saturated versus unsaturated lipids, and 3) The rate of membrane protein and membrane lipid production. These machineries form an intricate regulatory circuit that jointly controls membrane biogenesis. Currently, his laboratory is developing new approaches to isolate organelles

Emerging roles of the unfolded protein response in membrane homeostasis

Lipids and proteins shape the physicochemical properties of cellular membranes as a collective, thereby affecting membrane protein function and contributing to organelle identity. Eukaryotic cells face the challenge of maintaining the complex composition of several coexisting organelles. The molecular mechanisms underlying the homeostasis of subcellular membranes and their adaptation during stress are only now starting to emerge. We have studied three membrane property sensors of the endoplasmic reticulum (ER), namely OPI1, MGA2, and IRE1, each controlling a large cellular program impacting on the lipid metabolic network. OPI1 coordinates the production of membrane and storage lipids (Hofbauer et al., 2018), MGA2 regulates the production of unsaturated fatty acids required for membrane biogenesis (Covino et al., 2016), and IRE1 controls the unfolded protein response (UPR) to adjust the abundance and the folding capacity of the ER (Halbleib et al., 2017). Although these three machineries use remarkably distinct sensing mechanisms, they are functionally connected via the ER membrane and cooperate to maintain membrane homeostasis. Now, it is time to their cellular role by investigating their impact on the molecular composition of isolated organelles from stressed and unstressed cells. Based on a discussion of the sensing mechanisms and the integrative functions of the above-mentioned machineries, we propose that IRE1 and the unfolded protein response can sense the protein-to-lipid ratio in the ER membrane to warrant a balanced production of membrane proteins and lipids via the UPR.

[1] Covino R., Ballweg S., Stordeur C., Michaelis J.B., Puth K., Wernig F., Bahrami A., Ernst A.M., Hummer G., Ernst R. (2016) A Eukaryotic Sensor for Membrane Lipid Saturation. Molecular Cell 63:49-59

[2] Halbleib K., Pesek K., Covino, R., Hofbauer H.F., Wunnicke D., Hänelt I., Hummer, G., Ernst R. (2017) Activation of the Unfolded Protein Response by Lipid Bilayer Stress. Molecular Cell 67:673-684

[3] Hofbauer H.F., Gecht M. Fischer S.C., Sybert A., Frangakis A.S., Stelzer E.H.K., Covino R., Hummer G., Ernst R. (2018) The molecular recognition of phosphatidic acid by an amphipathic helix in Opi1. Journal of Cell Biology, 217:3109-3126

.

and to study their molecular compositions in order to generate a quantitative cellular map with subcellular resolution.

Argyris Papantonis

University of Göttingen
Göttingen, Germany

Monday
5:30 pm

Argyris Papantonis received his PhD on Biochemistry and Molecular Biology in 2008 by the University of Athens, Greece. He went on to study transcriptional organization as a postdoctoral researcher at the University of Oxford, where he later took up a Lecturer of Biochemistry position at University College. In 2013 he started his own lab on Chromatin Systems Biology at the University of Cologne, and in 2018 he was appointed as a Professor of Translational Epigenetic with the Medical Faculty of the University of Göttingen. His laboratory focuses on understanding the basic rules behind chromatin and transcriptional organization in 3D space and time with an emphasis on models of cellular senescence, ageing or pro inflammatory signaling

Impaired chromatin remodelling links misregulation of autophagy and DNA damage response to a rare progeroid syndrome

Members of the Chromodomain-Helicase-DNA binding (CHD) protein family are chromatin remodelers critically implicated in human pathologies. CHD6 is the least studied member of this family, and we were motivated to dissect its in-cell roles by discovering a mutated CHD6 allele in a patient suffering from the rare Hallermann-Streiff premature aging syndrome (HSS). We generated isogenic iPSC lines carrying (or not) this single point mutation in the CHD6 SANT/SLIDE domain, which allow studying HSS-relevant cell identities. Using these lines, we show for the first time that CHD6 binds the promoters of a cohort of autophagy and stress response genes across cell types. The mutation in CHD6 impairs its ability to recruit co-factors and regulate genes in response to DNA damage and autophagy stimulation, thus leading to accumulation of unresolved DNA damage burden. By combining genomics and functional assays, we describe a molecular mechanism underlying chromatin control of autophagic flux and genotoxic stress surveillance that is broadly applicable to human cell types and can explain HSS onset and progression.

Notes

Randy Hampton

University of California
San Diego, California, USA

Tuesday
9:30 am

Randolph Hampton was born in New York State just before cars evolved fins. He grew up in Scarsdale, NY, attended Ohio Wesleyan University, and by a circuitous route that included research at University of Michigan, running the office at Zingerman's Delicatessen, and doing field work in music and comedy, finally earned a Ph.D. in Biochemistry in the laboratory of Christian Raetz at University of Wisconsin-Madison and at Merck & Co. in Rahway, NJ. He then headed west to do a postdoctoral fellowship with Jasper Rine at U. C. Berkeley, in one of the few laboratories in the world dedicated to both yeast and dog genetics, where his work on regulated degradation of HMG-CoA reductase (in yeast, not dog) began. He next took a position in the Division of Biological Sciences at U. C. San Diego, where he remains to this day. His research interests include cellular management of misfolded proteins, sterol regulation and the cholesterol pathway, metabolism, and ubiquitin in health and disease. His pedagogical interests involve finding ways to communicate to tomorrow's physicians and biomedical professionals the critical differences between sickness and wellness, and the role of science in exploring that boundary. In addition, Randy (which is what everyone calls him except when in he is in trouble) is also deeply involved in the recovery movement, with 36 years clean and sober so far. His interests outside the academe include running, 5 string banjo, secular spirituality, and the complex synergies that govern the lives of us all. Most recently,

Mallosteric Misfolding and Rhomboidal Retrotranslocation: Lessons From Regulated ERAD

ERAD is highly selective, allowing destruction of proteins subtly perturbed from their native conformation. Biology has capitalized on this selectivity to regulate the stability of normal native proteins. Understanding how the cell harnesses the specificity of ERAD to accomplish protein regulation can provide new routes for experimental alteration of quality control, useful both for basic and therapeutic applications. We have been studying the regulated ERAD of Hmg2, an integral membrane protein by the HRD ERAD pathway. The sterol pathway molecule GGPP directly causes Hmg2 to undergo a reversible transition to a HRD pathway substrate. GGPP regulates Hmg2 ERAD at the mid nanomolar range, and antagonism can be observed with closely related molecules. This appears to be an example of ligand-dependent misfolding, which we refer to as *mallostery* in reference to the classic mode of ligand mediated regulation of protein function. We believe that this mode of regulation has broad and untapped applications in chemical biology and pharma.

A subject of great interest in ERAD is the exit mechanism(s) for retrotranslocation of ERAD substrates to the cytosol. Using a new self-ubiquitinating substrate we have recently discovered a dedicated route for removal of integral membrane ERAD clients, referred to a ERAD-M substrates. This route of dislocation affects both HRD and DOA substrates, and is mediated by the Dfm1 rhomboid protein, which had previously been discounted as a participant, by us and others. This dichotomy is resolved with the realization that *dfm1* nulls undergo rapid suppression that depends on Hrd1, and the current best candidate for lumenal ERAD-L retrotranslocation. We have discovered that the stress imposed by the dfm1Δ null state results in dramatic remodeling of the ERAD machinery and the creation of a new retrotranslocation route for ERAD-M substrates. I will report on the primary function of Dfm1 in retrotranslocation, the features of the Hrd1-mediated retrotranslocation of ERAD-M substrates, and the novel stress that allows this suppression to arise.

(Aug 2018) he got married for the first time, to his lovely now-wife Marrie (with 2 r's), making him a newlywed who is applying for Medicare. The message: life is full of surprises.

Gaia Pigino

MPI of Molecular Cell Biology and Genetics
Dresden, Germany

Tuesday
10:15 am

Dr. Gaia Pigino received her PhD in Evolutionary Biology from the University of Siena, Italy in 2007 for her studies on bio-indicators for contaminated soil. Electron microscopy (EM), one of the tools she used for this work, quickly became the central method for her research ever since. After her first postdoc in the EM Lab of the Department of Evolutionary Biology in Siena, she moved to ETH Zurich and the Paul Scherer Institute in Switzerland and was awarded an EMBO fellowship to continue her work on the cryo-EM structural analysis of ciliary components. In 2012, Dr. Pigino started her own lab at the Max Planck Institute of Molecular Cell Biology and Genetics (MPI-CBG) in Dresden, Germany. The work of her lab is centered around the question how complex cellular machines self-organize. Cilia remain the main focus, where the Pigino Lab investigates fundamental functional aspects of the dynamic process required for the assembly of the cilium. Learning about self-organization and functional implications of structural aspects naturally requires to combine and further the latest imaging technologies (3D cryo-EM, tomography, correlative light and electron microscopy (CLEM), and various light microscopy techniques) with tools from biochemistry, in vitro reconstitution, and genetic engineering.

Mechanistic understanding of cellular processes by cryo-EM

New cryo-EM technologies enable to investigate protein structures in the native physiological context of the cell. We use these technologies to study the self-organized assembly of cilia, ubiquitous organelles of eukaryotic cells.

Assembly of the cilium requires the rapid bidirectional intraflagellar transport (IFT) of building blocks to and from the site of assembly at its tip. This bidirectional transport is driven by the anterograde motor kinesin-2 and the retrograde motor dynein-1b, which are both bound to a large complex of 25 IFT adaptor proteins. We have recently developed a millisecond resolution 3D correlative light and electron microscopy (CLEM) approach to show that anterograde and retrograde IFT trains use separated microtubule tracks along the microtubule doublets of the cilium (Stepanek & Pigino, 2016). With this method at hand we showed that the spatial segregation of oppositely directed trains ensures a collision free transport in the cilium. However, it remained to be explained how competition between kinesin and dynein motors, which are both found on the same anterograde trains, is avoided. In bidirectional transport systems in the cell, other than IFT, the presence of opposing motors leads to periodic stalling and slowing of cargos moving along the microtubule. No such effect occurs in IFT. To address these questions, we take advantage of the most advanced technologies in cryo-electron tomography and sub-tomogram averaging. After obtaining the 3D structure of IFT train complexes in the cilia of intact Chlamydomonas cells, we showed that a tug-

Notes

of-war between kinesin-2 and dynein-1b is prevented by loading dynein-1b onto anterograde IFT trains in an inhibited conformation and by positioning it away from the microtubule track to prevent binding. These findings show how tightly coordinated structural changes mediate the behavior of such a complex cellular machine.

Christopher Vollmers

University of California
Santa Cruz, California, USA

Tuesday
11:30 am

Christopher Vollmers graduated with a M.S. in Biomedical Sciences from the University of Wuerzburg (Germany). He then pursued my Ph.D. through a shared program between the University of Heidelberg (Germany) and the Salk Institute for Biological Studies in La Jolla, CA. After graduating, he joined the lab of Stephen Quake at Stanford University as a postdoctoral fellow. In the Quake lab, he worked on developing genomic tools to analyze B cells on population and single cell levels. He now continues this work with a focus on long read sequencing technologies in his own lab which he started in 2014 in the Biomolecular Engineering Department at UCSC.

Sequencing for transcriptome analysis

Recent advances in sequencing technology have the potential to enable a new generation of tools for bulk and single cell transcriptome analysis. In particular, long-read sequencing makes it possible to, at high throughput, sequence full-length transcripts as opposed to sequencing transcript fragments using short-read based RNA-seq. I will outline the big potential of long-read sequencing technology for transcriptome analysis and annotation and talk about my lab's work on overcoming inherent limitations of this technology.

Notes

Leonie Ringrose

Humboldt University of Berlin
Berlin, Germany

Tuesday
12:15 pm

Leonie Ringrose is a British molecular biologist on a journey towards theoretical biology. She has been a professor at the IRI for Lifesciences at Humboldt University, Berlin since May 2015. From 2006 to 2014 she was a junior group leader at the IMBA (Institute of Molecular Biotechnology) in Vienna. In the time between her junior and senior positions she did a short sabbatical in the group of Martin Howard at the John Innes Centre, UK, to learn about theoretical approaches to epigenetics. Her interest in mathematical modelling and quantitative biology began during her PhD at the EMBL, Heidelberg (1992-1997), where she used modelling to investigate quantitative aspects of DNA recombination.

Her group works on epigenetic regulation by the Polycomb and Trithorax group proteins in Drosophila and mouse development. These proteins are highly conserved and essential to life, but many aspects of their function are enigmatic. Leonie believes that theoretical approaches, in combination with quantitative experiments, offer great potential for unlocking some of the mysteries of epigenetic regulation. We use a combination of quantitative live imaging, mathematical modeling, computational approaches and molecular and developmental biology to understand the interaction of the Polycomb and Trithorax proteins with their chromatin targets. We aim to unravel this fascinating epigenetic gene regulatory system in terms of the design, function and dynamic behaviour of its components. Our goal is to understand

Epigenetics meets mathematics: The fusion of experiment and theory brings insights beyond intuition

Epigenetic gene regulation is highly stable: epigenetic memory of gene expression states can persist over many cell generations and potentially for longer. However, epigenetic regulation is also flexible: genes that are subject to epigenetic regulation can respond dynamically to environmental and developmental signals. How can epigenetic regulation be both stable and flexible? I propose that the key lies in the highly dynamic nature of epigenetic systems. Over the last two decades it has become clear that the nucleus is an extraordinarily busy and noisy place: many proteins, including epigenetic regulators, are in constant motion, exchanging rapidly between chromatin bound and free states. Quantitative aspects of this motion are highly regulated. I propose that that to fully understand this regulation, epigenetics needs mathematics. We need "moving models" built of mathematical descriptions, which we can feed with measured values of quantities and mobilities of the components. A good model makes testable predictions that tell us whether our hypothesis makes sense. If it does not, we change the model. There has never been a better time to combine theoretical approaches with quantitative experiments. On the theoretical side, the last decade has seen a quiet revolution in the application of models built by physicists to the deep questions of epigenetics. On the experimental side, the advent of technologies that allow real time analysis at the single cell and single molecule level, together with those that enable targeted genome editing, allow precise perturbation and quantitative measurements at an unprecedented level.

It is time for epigenetics to meet mathematics. I will give examples from work in the field and in my own lab, of how the fusion of experiment and theory has brought fresh insights into epigenetic regulation that go beyond intuition.

Notes

how a system whose components are in constant flux can ensure both stability and flexibility of gene expression states.

Miki Ebisuya

EMBL
Barcelona, Spain

Tuesday
1:50 pm

Miki Ebisuya did her under graduate and PhD research at Kyoto University, Japan. After getting her PhD in 2008, she became a group leader at Kyoto University in 2009. Her lab moved to RIKEN in 2013, and then moved again to EMBL Barcelona, Spain, in 2018. The research interest of her group is reconstituting developmental mechanisms in cell culture.

Human time vs. Mouse time: *in vitro* segmentation clock as a model system

Different species have different tempos of development: larger animals tend to grow more slowly than smaller animals. My group has been trying to understand the molecular basis of this interspecies difference in developmental time, using the segmentation clock as a model system.

The segmentation clock is the oscillatory gene expressions that regulate the timing of somite formation from presomitic mesoderm (PSM) during embryogenesis. We have recently succeeded in inducing PSM from both human iPS cells and mouse ES cells, detecting the oscillation and traveling wave of segmentation clock in vitro. Interestingly, the oscillation period of human segmentation clock was 5-6 hours while that of mouse was 2-3 hours. Taking advantage of our in vitro system and simple mathematical models, we have been comparing the genome sequences and molecular processes of the segmentation clock between human and mouse to explain the interspecies difference in the oscillation period.

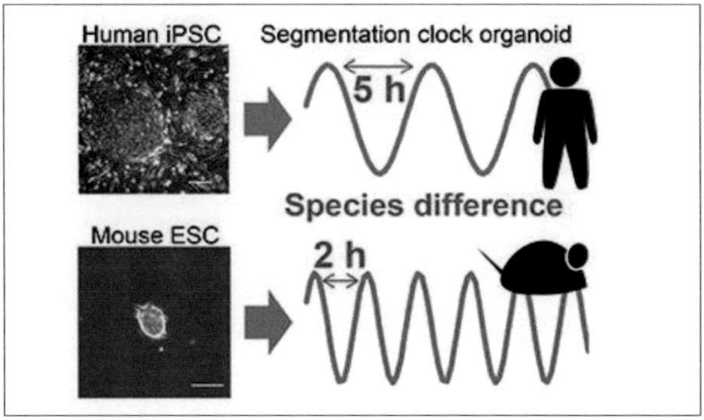

Notes

Needhi Bhalla

University of California
Santa Cruz, California, U.S.A.

Tuesday 2:35 pm

Needhi Bhalla received her undergraduate degree in biochemistry from Columbia College and her PhD in biochemistry from University of California, San Francisco. She performed her post-doctoral work at University of California, Berkeley and Lawrence Berkeley National Labs. She started her lab at University of California, Santa Cruz in 2008, where she studies chromosome structure and function in both meiosis and mitosis in *C. elegans*. She is particularly interested in how chromosome dynamics are monitored to ensure proper chromosome segregation during cell division. The work in her lab has demonstrated a surprising connection between the regulation of meiotic prophase events and the spindle checkpoint. In addition to her scientific research, she is interested in increasing diversity, equity and inclusivity in science, technology, engineering and mathematics (STEM) at all levels.

A tale of two checkpoints

How do cells ensure that the correct number of chromosomes is maintained with every cell division? Having an incorrect number of chromosomes, also called aneuploidy, is associated with cancer, birth defects, miscarriages and infertility, underscoring the importance of this question to human health. We combine genetic and biochemical approaches with high-resolution microscopy and cytological techniques in the nematode worm *C. elegans* to better understand how chromosomes are partitioned correctly during both sexual reproduction (meiosis) and development (mitosis). Currently, many of our studies focus on the conserved ATPase, PCH-2/TRIP13, which is required to monitor how homologous chromosomes interact in meiotic prophase as well as whether chromosomes are properly attached to the spindle during chromosome segregation. We are interested in determining whether common mechanisms underlie these diverse but essential chromosomal events.

Notes

Zoya Ignatova

University of Hamburg
Hamburg, Germany

Tuesday
3:55 pm

PhD with Volker Kasche within the Graduate School of TU Hamburg Harburg and University Hamburg. Postdoc with Lila Gierasch at the University of Massachusetts, Amherst, USA. Group Leader at MPI for Biochemistry, Martinsried. Tenure-track professor for Biochemistry at University of Potsdam and since 2014 full professor in Biochemistry and Molecular Biology at the University of Hamburg. 2016-18 Managing Director of Institute of Biochemistry and Molecular Biology. Prizes and awards: Heisenberg fellowship award; Lundbeck foundation, Denmark; Joliot professorship at ESPCI, Paris, France; Innovation award, MDI BioLab, Maine, USA; Hamburg teaching award; Claussen-Simon-Stiftung award for Mentorship.

Translational control: probing dimensionality beyond linear sequence of mRNA

The ribosome is a central molecular machine that translates the genetic information into a corresponding polypeptide in an mRNA template-directed manner. Thereby, the mRNA is not a mere messenger for translation of codons into amino acids but bears additional layers of information for folding, solubility and expression level of the encoded protein. We use cell-wide approaches, including quantification of translational components, probing mRNA secondary structure, mRNA abundance and translation efficiency by means of deep sequencing to extract features that are selected to control translation and influence protein folding and function.

Notes

Asifa Akhtar

MPI of Immunobiology and Epigenetics, Freiburg, Germany

Tuesday 4:40 pm

Asifa Akhtar obtained her bachelor's degree in Biology at University College London (UCL), UK, in 1993 and her Ph.D. in 1998 at the Imperial Cancer Research Fund in London, studying transcription regulation in Richard Treisman's laboratory. She continued in the field of chromatin regulation as a postdoctoral fellow at the European Molecular Biology Laboratory (EMBL), Heidelberg, Germany, and the Adolf Butenandt Institute, Munich, Germany, in Peter Becker's laboratory until 2001. From 2001, she led her own research as a group leader at the EMBL. Currently, Asifa Akhtar is a director at the Max Planck Institute of Immunobiology and Epigenetics in Freiburg, Germany, heading the Department of Chromatin Regulation. Her laboratory primarily studies chromatin and epigenetic mechanisms, especially focusing on the regulation by histone acetylation. In 2008, Asifa Akhtar received the European Life Science Organization (ELSO) award for significant contribution in the field and in 2013 she was elected as an EMBO member. In 2017, she received the prestigious Feldberg Prize.

Epigenetic regulation by histone acetylation

Our lab is studying the chromatin and epigenetic mechanisms regulated by histone acetylation using evolutionary conserved complexes associated with MOF, a MYST family of histone acetyltransferases. In flies and mammals, MOF is associated with the MSL and NSL complexes, which are important regulators of gene expression. In flies, the MSL complex is well known for regulation of the X chromosome by the process of dosage compensation, while the NSL complex regulates expression of house-keeping genes. In mammals, both complexes appear to be involved in regulating diverse cellular processes. The recent progress of our work will be presented.

Notes

Jen Heemstra

Emory University,
Atlanta, Georgia, USA

Wednesday 9:30 am

Jen received her B.S. in Chemistry from the University of California, Irvine, in 2000. At Irvine, she performed undergraduate research with Prof. James Nowick investigating the folding of synthetic beta-sheet mimics, which instilled in her a love of supramolecular chemistry. Jen then moved to the University of Illinois, Urbana-Champaign, where she completed her Ph.D. with Prof. Jeffrey Moore in 2005 studying the reactivity of pyridine-functionalized phenylene ethynylene cavitands. After a brief stint in industry as a medicinal chemist, she moved to Harvard University to pursue postdoctoral research with Prof. David Liu exploring mechanisms for templated nucleic acid synthesis. In 2010, Jen began her independent career in the Department of Chemistry at the University of Utah, and was promoted to Associate Professor with tenure in 2016. In 2017, Jen and her research group moved to the Department of Chemistry at Emory University. Research in the Heemstra lab is focused on harnessing the molecular recognition and self-assembly properties of nucleic acids for applications in biosensing and bioimaging. Outside of work, Jen enjoys spending time with her husband and two sons, as well as rock climbing, cycling, and running.

A chemical biology toolbox for specific RNA modification and capture

Nucleic acids are exquisitely adept at molecular recognition and self-assembly, enabling them to direct nearly all of the processes that make life possible. These capabilities have been fine-tuned by billions of years of evolution, and more recently, have been harnessed in the laboratory to enable the use of DNA and RNA for applications that are completely unrelated to their canonical biological roles. In our lab, we seek to use DNA and RNA for applications in biosensing and biomolecular imaging. We have developed RNA sequences that are capable of recognizing specific small molecule fluorophores and promoting covalent self-labeling with these fluorophores. We anticipate use of these self-labeling ribozymes for imaging of RNA in living cells. We have also demonstrated selective enrichment of inosine-containing RNAs using EndoV for non-covalent capture, and we anticipate this method will enable the identification of new sites of A-to-I editing in cells.

Notes

Tatjana Kleele

Laboratory of Experimental Biophysics, Lausanne, Switzerland

Wednesday 10:15 am

2015-today: Postdoc with Suliana Manley, Laboratory of Experimental Biophysics (EPFL, Switzerland)

- Live-cell super-resolution imaging of mitochondrial fission
- Physiological properties preceding mitochondrial fission at the single organelle level
- Development of a biophysical model for mitochondrial fission

2011-2014: PhD thesis with Thomas Misgeld, Institute of Neuronal Cell Biology (TU Munich)

- Development of an assay to study microtubule dynamics *in vivo* in the healthy and diseased nervous system
- *In vivo* microscopy techniques to study subcellular dynamics in mammalian neurons
- PhD program: Medical life science and technology

2008-2010: Research assistant with Bert Sakmann (Max-Planck-Institute of Neurobiology, Martinsried)

- Reconstruction of cortical neurons using bright-field mosaic scanning microscopy and semi- automated image processing

2004-2010: Studies in Biology (Neurobiology, Human Genetics and Organic Chemistry) at Ludwig- Maximilians-University Munich

Tatjana.Kleele@epfl.ch

Degradation and proliferation are outcomes of distinct types of mitochondrial division

"Mitochondria originated through the integration of an α-proteobacterium as a permanent organelle into a eukaryotic host cell. They became vital components of eukaryotic cells, executing critical functions, ranging from ATP production and Ca^{2+} handling to pro-apoptotic and reactive oxygen species signaling. Mitochondria are dynamic organelles constantly changing shape and size, depending on the metabolic demand of the tissue. The role of mitochondria in energy production makes them particularly vulnerable to reactive oxygen species, which can disturb protein structure and cause mtDNA mutations. Mitochondrial quality control mechanisms are counteracting this damage and regulate mitochondrial turnover: damaged mitochondria are selectively removed via mitophagy, and biogenesis of new mitochondria balances the removal of damaged mitochondria. Most of the molecular machinery regulating fission has been identified, but it remains obscure what distinguishes dysfunctional mitochondria from proliferating ones and how they are recognized by the division machinery.

Using live-cell super-resolution Structured Illumination Microscopy (SIM, iSIM) for ultra-fast multi-colour acquisition in Cos-7 cells and primary mouse cardiomyocytes, we analysed more than 2000 fission events. We discovered a key morphological determinant of the decision to proliferate or degrade: the positioning of the division site, giving rise to either "asymmetric" or "symmetric" division. By imaging fluorescent metabolic sensors, we find that asymmetric fissions show signs of stress and dysfunction and are followed by mitophagy. Symmetric fissions do not

Notes

Sjors Scheres

Medical Research Council
Laboratory of Molecular Biology,
Cambridge, United Kingdom

Wednesday 11:30 am

Sjors studied chemistry at Utrecht University, The Netherlands, where he also obtained his PhD in protein crystallography. He was a post-doc in the group of Jose-Maria Carazo in Madrid, before he started his group at the LMB in 2010. His main interest is currently in the development of new methods for high-resolution cryo-EM structure determination and in the cryo-EM structure determination of amyloid fibrils.

The atomic structures of Tau filaments from Alzheimer's and Pick's disease brains

The assembly of microtubule-associated protein tau into abundant filamentous inclusions underlies many neurodegenerative diseases called tauopathies. Tau inclusions display distinct neuroanatomical and cellular distributions between different tauopathies. Morphological and biochemical differences suggest that tau filaments adopt disease-specific molecular conformations. There are six tau isoforms - three isoforms with four microtubule-binding repeats each (4R tau) and three isoforms lacking the second repeat (3R tau). This gives rise to tauopathies with filaments composed of only 3R tau, only 4R tau or both 3R and 4R tau. Molecular conformers of filamentous tau may give rise to different neuropathological phenotypes, similar to prion strains, but the underlying structures are not known. Using electron cryo-microscopy (cryo-EM), we previously determined the structures of tau filaments from Alzheimer's disease, which contain both 3R and 4R tau. More recently, we also determined the structures of tau filaments from Pick's disease and show that they define a second tau protein fold consisting of residues K254-F378 of 3R tau, thereby proving the existence of molecular tau conformers. I will present both folds and discuss how they explain differences in isoform incorporation and phosphorylation observed between Alzheimer's disease and Pick's disease.

Notes

Max Cryle

EMBL Australia,
Monash University
Melbourne, Australia

Prof. Max Cryle is an NHMRC Career Development Fellow and EMBL Australia Group leader based in the Biomedicine Discovery Institute at Monash University. After obtaining his PhD in chemistry from the University of Queensland in 2006, he moved to the Max Planck Institute for Medical Research in Heidelberg as a Cross Disciplinary Fellow of the Human Frontiers Science Program. He was subsequently awarded funding from the German Research Foundation (Deutsche Forschungsgemeinschaft) to establish his own group to investigate glycopeptide antibiotic biosynthesis as part of the Emmy Noether program. His group works at the boundary of chemistry and biology, where they apply a multidisciplinary approach including synthetic chemistry, biochemistry, structural biology and enzyme catalysis. In 2016, he joined EMBL Australia to continue his research into understanding the biosynthesis of important natural antibiotics and developing new antimicrobial agents. His group has made a number of important breakthroughs in understanding how nature synthesises the glycopeptide antibiotics, which are clinically relevant and synthetically complex molecules. For this work, he was awarded the 2016 Otto Schmeil prize by the Heidelberg Academy of Arts and Sciences. Currently, his group is investigating the biosynthesis of several important antibiotics as well as investigating novel strategies and targets for antimicrobial development.

Understanding the biosynthesis of the glycopeptide antibiotics

The glycopeptide antibiotics (GPAs) are a structurally complex and medically important class of peptide natural products that include the clinical antibiotics vancomycin and teicoplanin. They contain a large number of non-proteinogenic amino acids and are produced by a linear non-ribosomal peptide synthetase (NRPS) machinery comprising seven modules. Furthermore, GPAs are extensively crosslinked late in their biosynthesis on the NRPS assembly line by the actions of a cascade of Cytochrome P450 enzymes, a process which contributes to the rigidity and structural complexity of these compounds. Due to the challenge of synthesising GPAs, biosynthesis remains the only means of accessing GPAs for clinical use, which makes understanding the biosynthesis of GPAs of key importance.

Notes

In this presentation, I will detail results from our studies into the NRPS machinery, the P450-catalysed cyclisation cascade and the interplay of these two important biosynthetic processes during GPA biosynthesis. This includes the characterisation of key enzymatic processes during NRPS-mediated peptide biosynthesis (chlorination, thioesterase activity and reconstitution of peptide synthesis) as well as the P450-mediated cyclisation cascade (substrate specificity of P450 enzymes and cascade reconstitution) from a number of different GPA biosynthesis pathways. Overall, our results demonstrate how selectivity during GPA biosynthesis is mediated through the careful orchestration of critical modification steps and interactions between the peptide-producing NRPS machinery and trans-modifying enzymes.

Margaret McCarthy

University of Maryland,
Baltimore, Maryland, USA

Wednesday
3:00 pm

Margaret McCarthy received a PhD from the Institute of Animal Behavior at Rutgers University, and postdoctoral training at Rockefeller University. She was an NRC Fellow at NIAAA and joined the faculty of the University of Maryland School of Medicine in 1993. She was a professor in the Department of Physiology before becoming the Chair of the Department of Pharmacology in 2011. She has received numerous awards and recognition for her research and for mentoring of graduate students. McCarthy has a long standing interest in the cellular and molecular mechanisms establishing sex differences in the brain. She uses a combined behavioral and mechanistic approach in the laboratory rat to understand normal brain development and how this process might go selectively awry in males versus females, discovering numerous novel signaling processes along the way.

Surprising origins of sex differences in the brain

Elucidating the cellular and molecular mechanisms by which sex differences are developmentally programmed into the brain has been a central goal of neuroendocrinology since the discipline was formed half a century ago. Neuroanatomical sex differences range from cell number, to phenotype, to dendritic morphology and synaptic patterning. Research emphasis has largely been on the intersection of steroids, neurotransmitters and growth/survival factors. Our laboratory has found a central role for a different source of neuromodulation, the neuroimmune system, and has discovered that at least two immune cells, microglia and mast cells, are critical partners in the process of brain masculinization. We further find that membrane derived signaling molecules, in particular the prostaglandins and endocannabinoids, are also essential drivers of the sexual differentiation process by modulating the activity of immune cells in the brain. Why males have higher levels of all of these is a mystery but may have its origins in the maternal immune system and its response to male fetuses during mammalian gestation.

Notes

Pavel Tomancak

MPI of Molecular Cell Biology and Genetics,
Dresden, Germany

Wednesday
3:45 pm

Pavel Tomancak studied Molecular Biology and Genetics at the Masaryk University in Brno, Czech Republic. He then did his PhD at the European Molecular Biology Laboratory in the field of *Drosophila* developmental genetics. During his post-doctoral time at the University of California in Berkeley at the laboratory of Gerald M. Rubin, he established image-based genome-scale resources for patterns of gene expression in *Drosophila* embryos. Since 2005 he leads an independent research group at the Max Planck Institute of Molecular Cell Biology and Genetics (MPI-CBG) in Dresden where he became senior research group leader in 2013.

His laboratory continues to study patterns of gene expression during development by combining molecular, imaging and image analysis techniques. The group has lead a significant technological development aiming towards more complete quantitative description of gene expression patterns using light sheet microscopy. The emphasis on open access resulted in establishment of major resources such as OpenSPIM and Fiji. The Tomancak lab is expanding the systematic analysis of gene expression patterns to other *Drosophila* tissues and employing the comparative approach in other invertebrate species to study the evolution of early development.

A new force awakens: comparative approach to tissue morphogenesis in insects

During gastrulation, physical forces reshape the simple embryonic tissue to form a complex body plan of multicellular organisms. These forces often cause large-scale asymmetric movements of the embryonic tissue. In many embryos, the tissue undergoing gastrulation movements is surrounded by a rigid protective shell. While it is well recognized that gastrulation movements depend on forces generated by tissue-intrinsic contractility, it is not known if interactions between the tissue and the protective shell provide additional forces that impact gastrulation. Our recent work has shown that a particular part of the blastoderm tissue of the red flour beetle *Tribolium castaneum* tightly adheres in a temporally coordinated manner to the vitelline envelope surrounding the embryo. This attachment generates an additional force that counteracts the tissue-intrinsic contractile forces to create asymmetric tissue movements. Furthermore, this localized attachment is mediated by a specific integrin, and its knock-down leads to a gastrulation phenotype consistent with complete loss of attachment. Moreover, analysis of another integrin in the fruit fly *Drosophila melanogaster* suggests that gastrulation in this organism also relies on adhesion between the blastoderm and the vitelline envelope. Together, our findings reveal a conserved mechanism whereby the spatiotemporal pattern of tissue adhesion to the vitelline envelope provides controllable counter-forces that shape gastrulation movements in insects. It also provides a new perspective on evolution of early gastrulation processes impacted by patterned contacts with the constraining extra-embryonic envelopes.

Notes

Leo James

Medical Research Council
Laboratory of Molecular Biology,
Cambridge, United Kingdom

Leo received his Ph.D. from Cambridge University in 2000, where he worked on antibody structure and design including the first humanised antibody CAMPATH. He post-doc'd in the labs of Prof. Dan Tawfik and Sir Greg Winter, investigating molecular mechanisms of antibody pathogenicity. In 2007, Leo established an independent group at the Laboratory of Molecular Biology in Cambridge. His lab studies intracellular host-pathogen interactions using a broad range of in vitro and in vivo techniques. In 2010, Leo discovered TRIM21, the most conserved and highest affinity antibody receptor in mammals and unique for its cytosolic expression. Since then, his lab has shown that TRIM21 prevents infection by intercepting viruses, bacteria and pathogenic proteins inside the cell and targeting them for rapid degradation. This work has also led to the development of 'TrimAway', a technique which exploits TRIM21 for the rapid and specific degradation of cellular proteins. Leo's lab is now working to define the molecular mechanisms that underpin TRIM21 activity to further develop this technology and understand how intracellular antibody immunity is regulated. Leo also investigates HIV post-fusion biology; recent work includes identifying the HIV capsid interface used to recruit cofactors for nuclear import and discovering dynamic pores in the capsid that are essential for HIV infection.

Epigenetic regulation by histone acetylation

TRIM21 is a recently discovered mammalian Fc receptor and E3 ubiquitin ligase expressed in the cytosol of all cells. The highest affinity IgG receptor in man, it mediates an intracellular humoral response against antibody-opsonized pathogens that invade the cytosol during infection. Upon detection, TRIM21 activates key immune transcription pathways to induce a potent antiviral state. Simultaneous with, and independent of, immune activation, TRIM21 recruits cellular degradation machinery, which catalyse the disassembly and destruction of cytosolic virions to prevent their replication. This rapid virion destruction also exposes the genomes of RNA and DNA viruses, allowing TRIM21 to potentiate the activity of RIG-I and cGAS to promote a rapid antiviral response. TRIM21 is effective against diverse viruses but also bacteria and proteopathic agents like tau. In my talk I will summarize key aspects of TRIM21 biology and focus on recent work showing how its inflammatory signaling is regulated and how we have repurposed its activity to perform targeted protein depletion. This latter technology, called 'Trim-Away', uses off-the-shelf antibodies to rapidly and acutely degrade cellular proteins in diverse cells.

Notes

Naama Barkai

Weizmann Institute of Science, Rehovot, Israel

Thursday 10:15 am

Naama Barkai was born and raised in Jerusalem. During her doctoral studies in the Physics Department of the Hebrew University, she became interested in biological computation systems and began using tools of statistical physics to study models of neural networks. Her interests then shifted to computation systems that function within cells, and she joined a group at Princeton University that was studying a bacterial sensing system. Her theoretical studies, and the experimental studies that followed, established this bacterial sensing pathway as a main model for Systems Biology studies. Barkai joined the Weizmann Institute in late 1999, just as the genomic revolution was getting underway. She was fortunate to be among the physics-trained biologists who were applying tools and concepts from the quantitative sciences to study how cellular computation systems are designed. This group was the founding core of the emergent field of Systems Biology. Barkai's work contributed to two subfields of Systems Biology: understanding the design principles of biological circuits on a relatively small scale, and understanding principles of gene expression at the genomic level. A central contribution of Barkai's work is the formulation and application of the robustness principle. Biological circuits, in contrast (perhaps) to man-made computation devices, work within what is inherently a highly noisy environment, as manifested in different dimensions. The robustness principle suggests that the biological circuits selected by evolution are robust; that is,

Gene transcription as a limiting factor in protein production and cell growth

Cell growth is driven by the synthesis of proteins, genes and other cellular components. Defining processes that limit biosynthesis rates is fundamental for understanding determinants of cell physiology. We address this challenge by combining mathematical modeling with the analysis of cells forced to express high levels of inert proteins as a tool to examine the consequences of increasing biosynthetic demands. I will describe the parameters which define the maximum possible values by which cells can translate proteins or transcribe genes, and present results suggesting that cells work close to these 'hard' limits. I will them discuss the possible consequences of these 'hard' limits on cell growth rate and cell size, the interplay between these two parameters, and the potential evolutionary benefits of maximizing transcription rates.

they perform their function reliably in a noisy environment, showing minimal dependency on the kinetics of quantitative parameters. Over the years, Barkai has applied this principle to reveal design principles and operational mechanisms of multiple circuits that work in diverse contexts. In particular, her work has revealed mechanisms that function during multicellular development to pattern the body plan.

Barkai is a faculty member in the Department of Molecular Genetics at the Weizmann institute. She has received several awards in recognition of her work, including the EMBO award for women in science and the Rothschild Prize.

James Williamson

The Scripps Research Institute, California, USA

Thursday 2:30 pm

Dr. James Williamson is the Executive Vice President of Research and Academic Affairs at The Scripps Research Institute, as well as a Professor in the Departments of Chemistry and Integrative Structural & Computational Biology. His research involves the study of RNA structure, RNA-protein interactions and RNA-ligand interactions using biochemistry, biophysics and structural biology approaches.

Among his many early scientific career accomplishments, Dr. Williamson was named a Searle Scholar, a Rita Allen Scholar, a Camille Dreyfus Teacher-Scholar, a Sloan Research Fellow and was awarded a Jane Coffin Childs Postdoctoral Fellowship. His studies have appeared in more than 200 peer-reviewed publications and in 2010, he was elected to the American Academy of Arts and Sciences. He currently serves on editorial boards for a number of top scientific journals and professional societies.

At Scripps Research, Dr. Williamson continues to successfully execute multiple leadership roles, especially those encompassing academic planning and training. In 2001, he was named Associate Dean for the Chemistry program and from 2008-2017, he was the Dean of Graduate and Postdoctoral Studies for the Institute's nationally ranked graduate program. He was named Vice President of Academic Affairs in 2015 and appointed Executive Vice President in 2017.

Ribosome Assembly in Bacteria

Ribosomes are responsible for all protein synthesis in cells, and the process of assembling ribosomes accounts for one third of the energy budget for rapidly growing bacteria. The assembly process involves coordination of synthesis of three large ribosomal RNAs and fifty ribosomal proteins in an incredibly efficient process that requires about two minutes in cells. Assembly is guided by over fifty assembly factors that are chaperones and modification enzymes, but we have a basic and emerging understanding of the steps involved in making a ribosome. Our laboratory uses a wide variety of biophysical methods to characterize the structure and dynamics of assembling ribosomes in vitro, and in cells. Quantitative mass spectrometry reveals the composition of intermediates, electron microscopy reveals the heterogeneous distribution of assembly intermediates, and single molecule fluorescence reveals the dynamics of structural transitions that occur during assembly. The overall picture that is emerging is that there are parallel pathways for installation of various structural units, but that certain events must happen in a particular order. The parallel and sequential nature of assembly appears to be an optimal balance for efficient and accurate assembly. This talk will describe recent advances in our understanding of ribosome assembly in bacteria, focusing on general principles that are emerging about the extensive RNA folding that is at the heart of the process.

Dr. Williamson earned his doctorate in chemistry at Stanford University in 1988. Following postdoctoral work at the University of Colorado, he joined the faculty of the Chemistry department at the Massachusetts Institute of Technology, where he attained the rank of associate professor with tenure. He accepted an appointment to The Scripps Research Institute as a professor in 1998.

Michael Levine

Princeton University,
Princeton, New Jersey, USA

Thursday 3:15 pm

Professional Experience:

July 2015: Director, Lewis-Sigler Institute for Integrative Genomics, Princeton University.

July 2015: Anthony B. Evnin Professor, Dept Molecular Biology, Princeton Univ.

July 2012-June 2015: Chair, Chancellor's Advisory Council for Biology, UC Berkeley

July 2007-June 2011: Head, Division of Genetics, Genomics, & Development, Dept MCB, UCB

July 2002-June 2015: Co-Director, Center for Integrative Genomics

July 2001-June: 2002 Associate Director, Functional Genomics, Joint Genome Institute, DOE, Walnut Creek, CA

Sept. 1999-Aug. 2000: Visiting Professor, Zoological Institute, University of Zürich, Switzerland

July 1996 - present: Professor, Department of Molecular and Cell Biology, UC Berkeley

Jan. 1991-June 1996: Professor, Department of Biology, University of California, San Diego

July 1988-Dec. 1990: Professor, Department of Biological Sciences, Columbia University, New York (NY)

July 1986-June 1988: Associate Professor, Department of Biological Sciences, Columbia, University, NY

July 1984-June 1986: Assistant Professor, Department of Biological Sciences, Columbia University, NY

Enhancer-promoter communication in living *Drosophila* embryos

Transcriptional enhancers are short segments of DNA (~100 bp to 1 kb) that switch genes on and off in response to intrinsic and external cellular signals. Whole-genome assays suggest that the human genome contains ~400,000 enhancers (~20 per gene). There is emerging evidence that sequence polymorphisms in enhancer DNAs are a major source of population diversity and predilection to disease. My lab uses quantitative live imaging methods to visualize the activities of enhancers that control the development of the early *Drosophila* embryo. I will discuss transcriptional bursts, coordinate activation of linked genes, and transvection, whereby an enhancer located on one chromosome can activate the expression of a transcription unit located on the other homologue.

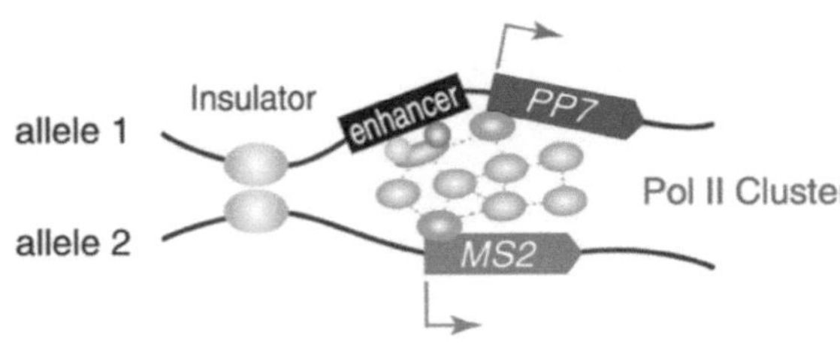

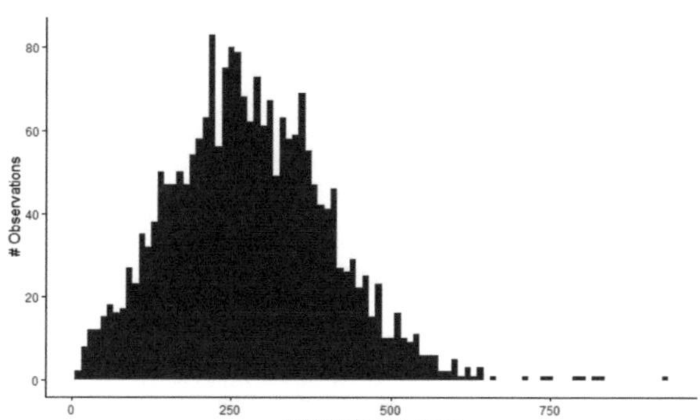

Oct. 1983-June 1984: Postdoc, Department of Biochemistry, University of California, Berkeley; Dr. G.M. Rubin, Supervisor

Apr. 1982-Sept. 1983: Postdoc, Department of Cell Biology, Universität Basel; Dr. W. Gehring, Supervisor

Education:

B.A., 1976 University of California, Berkeley, Department of Genetics

Ph.D., 1981 Yale University, Department of Molecular Biophysics & Biochemistry

Honors:

1976 Phi Beta Kappa; Departmental Citation in Genetics. UC Berkeley

1982-1984 Jane Coffin Childs Postdoctoral Fellow

1985-1988 Searle Scholars Fellow

1985-1987 Alfred P. Sloan Research Fellow

1996 Award in Molecular Biology, National Academy of Sciences

1996 Elected to the American Academy of Arts & Sciences

1998 Elected to the National Academy of Sciences

1999 Harvey Lecture

2002 F. Williams Chair in Biology, UC Berkeley

2006 Gavin Borden Fellow, Cold Spring Harbor Labs

2009 Faculty Research Lecturer, UC Berkeley; Wilbur Cross Medal, Yale University

2010 Einstein Professor, Chinese Academy of Sciences

2012 President, Society for Developmental Biology

2015 Conklin Medal, Society for Developmental Biology

2017 Elected, Associate Member EMBO

PANEL DISCUSSION

Panel Discussion

The panel discussion has been organized every year in Horizons since its conception. It provides a channel for lively interactions in a less formal setting. Typically, we have five guests from diverse disciplines on the panel taking questions from a moderator and from the audience. The idea is to illuminate different aspects of the topic from various perspectives. Horizons strives to have a diverse range of speakers to guide us with their unique experiences. For the 16th Horizons in Molecular Biology symposium, the following speakers will join the panel:

1. Needhi Bhalla
2. Michael Levine
3. Andreas Laustsen
4. Christine Mieck
5. Katrin Wodzicki

Being a scientist: Job or Passion?

Many of us were driven to pursue science because we were fascinated by the marvels and beauty of the natural world around us. But what keeps us in the race? Is it purely our quest to answer challenging scientific questions? Or is it a more career oriented desire to succeed in academic life? Or should we think of science more as a 9-5 job? The creative process of science is rarely a straightforward "Eureka" moment. Obstacles faced by scientists include long work-hours, pressure to publish and get grants and the need for success at each career stage. The accompanying stress can affect social and personal life and in extreme cases can lead to mental health issues.

In the panel discussion, we will discuss the ebb and flow of academia – from the exciting quest for knowledge to the harsh reality – emphasizing some aspects of the vast landscape of what it means to be a scientist. Topics chosen for this year's discussion are:

- Being a scientist, Job or Passion?
- How critical is the choice of scientific project and group in which to pursue it? How can a productive PI-student relationship be established? How can the power and responsibility of a PI be used to benefit this relationship?
- Work-Life Balance: Who decides what is reasonable? When things go wrong what help is available and from whom?
- Communication in science. What forms does it take and how important is it?

This year, Emeritus **Professor Mary Osborn** has once again kindly agreed to moderate the discussion. She will be joint in the moderation by three students from the organizers team of the symposium: Ida Jentoft, Tarana Nigam, and Panagiotis Poulis.

Furthermore, we would like to announce the participation of **Katrin Wodzicki**, who is the head of the *Human Resources and Organization Development Unit* of the University of Göttingen.

Mary Osborn

Mary Osborn received her BA in Physics from the University of Cambridge and a PhD in Biophysics from Pennsylvania State University. She was a post-doc in Jim Watson's lab at Harvard University and a staff scientist in Sydney Brenner and Francis Crick's division at the MRC Laboratory of Molecular Biology in Cambridge, U.K. She then joined the staff at Cold Spring Harbor Laboratory. In 1975, she moved to the Max Planck Institute for Biophysical Chemistry in Göttingen and in 1989 was appointed an honorary professor at the University of Göttingen.

Her research has focused on different areas in the life sciences. Her 1964 paper with Klaus Weber, on the use of SDS gels to determine molecular weight is a Citation Classic. Her work on the use of antibodies in immunofluorescence microscopy allowed the definition of the arrangements of microfilaments, microtubules and intermediate filaments in cells and tissues. The 1978 demonstration of a 1:1 correspondence between microtubules viewed in immunofluorescence and electron microscopy in the same cell was a major breakthrough as was the work showing that antibodies specific for the different types of intermediate filament proteins are powerful reagents in human tumor diagnosis.

She was a trustee of the Swedish Foundation on the Environment, MISTRA, and has chaired the Scientific Advisory Boards of the European Molecular Biology Laboratory (EMBL) in Heidelberg and the Pette Institute in Hamburg. She also chaired the Cell Biology Section of Academia Europaea. She was a member of the European Strategy Forum on Research Infrastructures (ESFRI) Group. She has participated in several juries concerned with awarding substantial grants to young scientists including the BioFuture program and the Helmholtz Young Investigator Program and was a member of the Helmholtz Senate. She was President of the International Union biochemistry and Molecular Biology from 2003 to 2006. She chaired the ETAN expert working group appointed by the European Commission that produced an influential report on Women in Science published in 2000.

She has been an EMBO member since 1978 and she holds an honorary doctorate from the Pomerian Medical Academy in Sczeczin, Poland.

She has been awarded the Meyenburg Prize and the L'Oreal/UNESCO Prize for her work on the cytoskeleton and the use of antibodies in tumor diagnosis. Her research and her efforts for Women in Science, were honored by the award of the Bundesverdienstkreuz (1st class) in 2014.

Katrin Wodzicki

Katrin is head of the HR and organization development unit of the University of Göttingen since 2017, responsible for the development of scientific, technical and administrative staff as well as for the implementation of the tenure track programme and the further development of the structure of academic career paths in general. Katrin studied Psychology at the University of Jena, and completed her Ph.D. at the University of Zurich, followed by five years of postdoc at the Leibniz-Knowledge Media Research Center (KMRC) in Tübingen. During her time in Tübingen, she took part in the think tank of the "Stiftung Neue Verantwortung" in Berlin, which aimed at encouraging an intersectional dialogue between science, business, politics, and society. Moreover, she is one of the founders of the group blog wissensdialoge. de on which we discuss and report about psychological research relevant for practitioners in knowledge management and organizational learning. From 2012 to 2016, she established the GGNB Career Service for postdocs and late-stage PhDs (continued as GAUSS Career Service) in Göttingen.

Katrin has always liked pushing the development of others - as of herself - forward. Everyone has specific abilities, skills, and talents. And there are numerous career options within and outside academia. Finding the right place to make use of one's own talents is challenging whilst at the same time exciting. By taking part in the panel discussion, she aims at providing useful information, answer questions and support informed individual decision-making. You can follow her on Twitter *(@KatrinW)*.

POSTER ABSTRACTS

1

Aydın Banu

The effect of epidermal growth factor and cholinergic receptors agonist / antagonists together with anticancer drug 5-FU in hepatocelluler carcinoma cell line

Küçük Onat[2], Tanrıverdi Bademci Ayşe Melek[1], Aydın Banu[1], Cabadak Hülya[1]

[1)] Department of Biophysics, Marmara University, Medical School, Istanbul, Turkey;
[2)] Student in Marmara University, Medical School, Istanbul, Turkey

---

2

Abhishek Upadhyay

An Inactivation Switch Enables Rhythms in a Neurospora Clock Model

Abhishek Upadhyay[1], Michael Brunner[2], and Hanspeter Herzel[1]

[1)] Institute for Theoretical Biology, Charité, Universitätsmedizin Berlin and Humboldt University of Berlin, Philippstr. 13, 10115 Berlin, Germany
[2)] Biochemistry Center, University of Heidelberg, Im Neuenheimer Feld 328, 69120 Heidelberg, Germany; *michael.brunner@bzh.uni-heidelberg.de*

3

Sarvinoz Atakulova

The impact of ABCB1, SLCO1B1, CYP3A4/5 polymorphisms on efficacy of ticagrelor in patients with acute coronary syndrome

Sarvinoz Atakulova[1], Karin Mirzaev[2], Egor Torbenkov[2], Dmitry Sychev[2], Kristina Ryzhikova[2]

[1)] I.M. Sechenov First Moscow State Medical University of the Ministry of Health of the Russian Federation (Sechenov University), Moscow, Russia
[2)] Russian Medical Academy of Continious Professional Education of the Ministry of Health of the Russian Federation, Moscow, Russia

---

4

Oleh Rymarenko

Composition and properties of XPO-1 dependent nuclear export signals

Oleh Rymarenko, Dirk Goerlich

MPI-BPC, Göttingen

5

Fiene Daniel

Analysis of mitochondrial homeostasis and damage via femtosecond laser-based nanosurgery

F. Daniel[1,2,3], D. Müller[1,2,3], D. Theidel[1,2,3], D. Hagenah[1,2,3], A. Heisterkamp[1,2,3], S. Kalies[1,2,3]

[1)] Institut of Quantum Optics, Gottfried Wilhelm Leibniz University Hannover, Germany
[2)] Cluster of Excellence REBIRTH, Hannover, Germany
[3)] Lower Saxony Centre for Biomedical Engineering, Implant Research, and Development (NIFE), Hannover, Germany

---

6

Maryam Mehdipour

How could semen microscopic factors use as a predictable item for Sperm DNA Fragmentation Index in IVF laboratories?

Maryam Mehdipour[1], Mahsa Afrough[1], Mahmoud Hashemitabar[1, 2]

[1)] Inferetility research and Treatment center, Khuzestan ACECR, [2)] Cellular and Molecular Research Center, School of Pharmacy, Ahvaz Jundishapur University of Medical Sciences, Ahvaz.

7

Anjana Ghelani

Next-generation Sequencing based Microbiome Analysis of Oxygen Minimum Zone of Arabian Sea Coast

Anjana Ghelani, R. K. Patel

Department of Life Sciences, Hemchandracharya North Gujarat University, Patan, India

---

8

Cesar Cordero Gomez

Generation of myelin specific hybridoma cell lines of preserved specificity but with different isotypes using a CRISPR-cas9 approach

Cordero Gomez C., Lodygin D., Faiß L., Strauß J., Lühder F., and Flügel A.

Universitätsmedizin Göttingen, Germany

9

Dominik Müller

Mechanistic investigation of sarcomeric elements in cardiomyocytes via femtosecond laser-based nanosurgery

D. Müller[1,2,3], F. Daniel[1,2,3], D. Theidel[1,2,3], S. Biswanath[2,4], R. Zweigerdt[2,4], A. Heisterkamp[1,2,3], S. Kalies[1,2,3]

[1)] Institut of Quantum Optics, Gottfried Wilhelm Leibniz University Hannover, Germany; [2)] Cluster of Excellence REBIRTH, Hannover, Germany; [3)] Lower Saxony Centre for Biomedical Engineering, Implant Research and Development (NIFE), Hannover, Germany; [4)] Leibniz Research Laboratories for Biotechnology and Artificial Organs (LEBAO), Department of Cardiac, Thoracic, Transplantation and Vascular Surgery (HTTG), Hannover Medical, School, Hannover, Germany

---

10

Eman Abbas

Chromatin Remodeling BAF Complexes Control Oligodendrocytes Development in the Embryonic Telencephalon

Eman Abbas[1], Kamila Kiszka[1], Linh Pham[1], Cemil Kerimoglu[3], Andre Fischer[3], Jochen F. Staiger[1,2] and Tran C. Tuoc[1,2]

[1)] Institute for Neuroanatomy, University Medical Center, Georg-August University Göttingen, Germany; [2)] Center for Nanoscale Microscopy and Molecular Physiology of the Brain (CNMPB), Göttingen, Germany; [3)] German Center for Neurodegenerative Diseases, Göttingen, Germany

---

11

Francesca Lugarini

Initiation of Autophagy signalling by HORMA domain proteins

Francesca Lugarini, Faesen A.C.

Dept. Signal Dynamics, Max Planck Institute for Biophysical Chemistry, Göttingen, Germany

---

12

Fouzia Qamar

Bioefficacy of selected Isolates against Insects Pests

Fouzia Qamar[1], Sumaira Anwar[2], Imran Sajid[2]

[1)] Lahore Garrison University, Lahore, Pakistan; [2)] Microbiology and Molecular Genetics Department, Punjab University, Quaid e Azam Campus, Lahore, Pakistan

---

13

Flávia Rezende

Renal Nox4 contributes to systemic redox homeostasis by controlling glutathione, methionine, cysteine and folate metabolism

Flávia Rezende[1,2], Luciana Hannibal[3], Maria Pires Pacheco[4], Tamara Jean Rita Bintener[4], Pedro Malacarne[1,2], Niklas Müller[1,2], Thomas Sauter[4], Katrin Schröder[1,2], Ralf P. Brandes[1,2]

[1)] Institute for Cardiovascular Physiology, Goethe-University, Frankfurt am Main, Germany; [2)] German Center of Cardiovascular Research (DZHK) (Partner site Rhein-Main), Frankfurt am Main, Germany; [3)] Centre for Paediatric and Adolescent Medicine, Medical Center, University of Freiburg, Germany; [4)] Life Sciences Research Unit, Université du Luxembourg, Luxembourg

---

14

Franziska Grüner

Identification of B Cell Receptor Regulated MicroRNAs in a Model of Aggressive B Cell Lymphoma

Franziska Grüner, Claudia Pommerenke, Hans G. Drexler, Sonja Eberth

Department of Human and Animal Cell Lines, Leibniz-Institute DSMZ-German Collection of Microorganisms and Cell Cultures, Braunschweig (Germany)

15

Nicole Horn

Investigating the role of mature microRNAs identified in the nucleus of malignant B cell lymphoma cell lines

Nicole Horn[1], Jéssica Arribas Arranz[2], Hans G. Drexler[1], Sonja Eberth[1]

[1)] Department of Human and Animal Cell Lines, Leibniz-Institute DSMZ-German Collection of Microorganisms and Cell Cultures, Braunschweig (Germany); [2)] Department of Cellular Biology, Physiology and Immunolgy, Autonomous University of Barcelona (Spain).

16

Yaroslav Faletrov

Novel Steroids with Artifical Side Chains as Substrates for Mycobacterial Enzymes

Yaroslav Faletrov[1], Renata Plocinska[2], Anna Brzostek[2], Qing Qing Lin[1], Aleksandra Falchevskaya[1], Matvey Horetsky[1], Agnieszka Wojtkielewicz[3], Jacek Morzycki[3], Vladimir Shkumatov[1], Jarosław Dziadek[2]

1) Research Institute for Physical and Chemical Problems, Belarusian State University, Minsk, Belarus, Faculty of Chemistry, Belarusian State University, Minsk, Belarus; *(e-mail: yaroslav82@tut.by)*; [2)] Institute of Medical Biology, Polish Academy of Sciences, Lodz, Poland; [3)] Institute of Chemistry, University of Białystok, Białystok, Poland.

17

Nadine Brandes

Lineage tracing analyses reveal the potential of CD^{4+} stem cells in skin homeostasis and regeneration

Nadine Brandes, PD Dr. Anja Uhmann

Institute of Human Genetics, University Medical Center Göttingen

---

18

Zuzana Kadlecova

Temporal control of endocytosis by phosphorylation of the main endocytic adaptor protein AP2

Zuzana Kadlecova, et al.

David Owen Lab, CIMR, University of Cambridge, Cambridge, United Kingdom

---

19

Ivana Borovská

Mutations in RNA Recognition Motifs of PUF60 and U2AF65 Occuring in Cancer and their Impact on 3' Splice Site Selection

Ivana Borovská[2], Jana Královičová[1,2], Monika Kubíčková[3], Peter J. Lukavsky[3], Igor Vořechovský[1]

1) University of Southampton, Faculty of Medicine, Southampton SO16 6YD, United Kingdom; 2) Slovak Academy of Sciences, Institute of Molecular Physiology and Genetics, 840 05 Bratislava, Slovak Republic; 3) Masaryk University, CEITEC, 625 00 Brno, Czech Republic

---

20

Ivonne M Bartsch

Bioinformatic Tools for Peptides Design with Phagocytic Activity for Pharmacological Porpuses

Ivonne M Bartsch[1,2], Constanza Cárdenas C[1], Fanny Guzmán Q[1], and Sergio Marshall G[2]

1) Peptides design and synthesis Laboratory; 2) Genetic and Molecular Immunology Laboratory Pontifical Catholic University of Valparaíso, Chile

---

21

João Nuno da Rocha Oliveira-Gomes

Metabolomics of Muscarin – Clinical Targets and Toxicological Effects

João Nuno da Rocha Oliveira-Gomes[1,2*]

1) Department of Public Health and Forensic Sciences, and Medical Education, Faculty of Medicine, University of Porto, Porto, Portugal; 2) Department of Surgery and Physiology, Faculty of Medicine, University of Porto, Porto, Portugal; *Email: jonugos@gmail.com

---

22

Joachim Maier

NMR studies on peptide-protein interactions contributing to phase separation

Joachim Maier[1], Daniel Sieme[1], Leo Wong[1], Arshiya Bhatt[2],Jürgen Wienands[2], and Christian Griesinger[1]

[1)] Department of NMR-based Structural Biology, Max Planck Institute for Biophysical Chemistry, Am Fassberg 11, 37077 Göttingen, Germany;
[2)] Institute of Cellular and Molecular Immunology, Georg August University of Göttingen, Humboldtallee 34, 37073 Göttingen, Germany

---

23

Samar Mamdouh Shawki

Liraglutide improves 3-NP- induced Huntington's disease: Impact on BDNF/TrkB/PI3K/Akt/GSK-3β/β-catenin/CREB trajectory and miR-130a

Shawki SM[1], Saad MA[2,3], Rahmo RM[4], and El-Abhar HS[2,5]

[1)] Dept. of Pharmacology and Toxicology, Faculty of Pharmacy, Misr International University, Cairo, Egypt; [2)] Dept. of Pharmacology and Toxicology, Faculty of Pharmacy, Cairo University, Cairo, Egypt; [3)] School of Pharmacy, Newgiza University, Cairo, Egypt; [4)] Department of Pharmacology and Toxicology, Faculty of Pharmacy, Misr International University, Cairo, Egypt; [5)] Dept. of Pharmacology and Toxicology and Biochemistry, Faculty of Pharmacy, Future University in Egypt, Cairo, Egypt.

24

Lubos Janotka

Development of Resistance to HMAS through Apoptosis Regulation

Lubos Janotka[1], Kristina Simonicová[1], Helena Kavcova[1], Zdena Sulova[1], Albert Breier[1,2], and Lucia Messingerova[1,2]

[1)] Department of Biochemistry and Cytochemistry, Institute of Molecular Physiology and Genetics, Centre of Biosciences, Slovak Academy of Sciences, Bratislava, Slovak Republic;
[2)] Institute of Biochemistry and Microbiology, Faculty of Chemical and Food Technology, Slovak University of Technology, Bratislava, Slovak Republic

---

25

Priscilla González

Effect of overexpression of the global transcription factor Xbp1 on the specific production and glycosylation patterns of erythropoietin produced by CHO cells

Priscilla González, Yesenia Latorrea, Shawal Spencer[2], Natascha Gödeckeb, Hansjörg Hauser[2], Dagmar Wirth[2], María Carmen Molina[3], Julio Berriosa, Raquel Montesinos[4], and Claudia Altamirano[1,5]

[1)] Biochemistry Engineering School, PUCV, Valparaíso, Chile; [2)] MSYS, HZI, Braunschweig, Germany; [3)] Inmunobiotechnology Institute, UChile, Santiago, Chile; [4)] Physiopathology Department, UDEC, Concepción, Chile; [5)] CREAS, Valparaíso, Chile.

26

Rashi Goel

New Approaches for Studying Neurotransmitter Transporters

Rashi Goel[1], Reinhard Jahn[1]

[1)] Department of Neurobiology, Max Planck Institute for Biophysical Chemistry, 37077, Göttingen, Germany

---

27

Rima Siauciunaite

Multiple clocks regulate amino acid levels in zebrafish cells

Rima Siauciunaite, Nicholas S. Foulkes

Institute of Toxicology and Genetics (ITG) Karlsruhe Institute of Technology, Hermann-von-Helmholtz-Platz 1, 76344 Eggenstein-Leopoldshafen, Germany

---

28

Georgy Selihanov

Crystallization in lipid mesophase can induce changes in crystal structure

Georgy Selihanov[1*], Tatiana Fufina[2], Lyudmila Vasilieva[2], and Azat Gabdulkhakov[1]

[1)] Institute of Protein Research, Russian Academy of Sciences, Pushchino, Russia;
[2)] Institute of Basic Problems of Biology, Russian Academy of Sciences, Pushchino, Russia

---

29

Mariam Ahsan

Serum Osteopontin, as a Biomarker Predicting the Magnitude of Disease Severity & Progression to Hepatocellular Carcinoma in Hepatitis C Virus Infected Patients in Pakistan

Mariam Ahsan[1,3],Talat Roome[1], Anam Razzak[1], Zahid Azam[2], and Mehreen Lateef[3]

[1)] Molecular Pathology, Department of Pathology, Dow Diagnostics Reference and Research Laboratory , Dow International Medical College, Dow University of Health & Sciences, Pakistan;
[2)] National Institute of Liver & GI diseases (NILGID), Dow University of Health & Sciences, Pakistan;
[3)] Multidisciplinary Research Lab, Bahria University Medical & Dental College, Pakistan;
Corresponding Author: Ms. Mariam, Email: *mariamnaz786@gmail.com*

30

Shaymaa Abdulmalek

Molecular Investigation of Natural Oil with Antidiabetic Drugs on Brain Insulin Signaling Pathway in Diabetic Rats

Shaymaa Abdulmalek

Department of Biochemistry, Faculty of Science, Alexandria University, Egypt

---

31

Yulia A. Sidorova

Small molecules mimicking biological activity of glial cell line-derived neurotrophic factor family ligands to target neurodegeneration.

Yulia A. Sidorova, Mart Saarma.

University of Helsinki, Faculty of Medicine, Finland

32

Sopiko Kurdadze

Prevalence of High Oncogenic Risk HPV Genotypes 16, 31, and 33 in Georgian Women with Cervical Intraepithelial Neoplasia Grade 1

Sopiko Kurdadze[1], Sahar Abd Elmogheth Madani[1], Keti Manjgaladze[2], Giorgi Kharebava[1]

[1)] Allele Ltd, Tbilisi, Georgia, [2)] XXX, Tbilisi, Georgia;
Corresponding author: Giorgi Kharebava, PhD, Allele Ltd, Lisi Lake #4, Tbilisi, 0186, Georgia, email: *kharebava@gmail.com*

33

Victor Tobiasson

The structure of the mitochondrial ribosome from T. thermophila

Victor Tobiasson, Alexey Amunts

Stockholm University, Stockholm, Sweden

34

Svetlana Ohiienko and Anastasiya Bondar

Cu-Induced Liver Fibrosis Affects the Primary Culture of Rats Bone Marrow Cells

S.L. Ohiienko, A.Yu. Bondar

Research Institute of Biology, V.N.Karazin Kharkov National University, 4 pl. Svobody, Kharkov 61022, Ukraine, e-mail: *ohiienko.svetlana@gmail.com*

35

Moritz W.R. Willmer

Ribosome recycling in a natural translation termination context

Moritz W.R. Willmer, Frank Peske, and Marina V. Rodnina

Department of Physical Biochemistry, Max Planck Institute for Biophysical Chemistry (MPI-BPC), 37077, Göttingen, Germany

---

36

Victor Hugo Canela

Laser microdissection of human calcium oxalate monohydrate stones enables structurally specific exploration of the stone proteome

Victor Hugo Canela, Sharon B. Bledsoe, James E. Lingeman, Tarek M. Ashkar, and James C. Williams Jr.

Department of Anatomy and Cell Biology, Indiana University School of Medicine, Indianapolis, Indiana

---

37

Murali Aarthy

Investigations on the mechanistic insights of wild and mutant E6 Oncoprotein of HPV and its interaction with the tumor suppressor protein

Murali Aarthy, Sanjeev Kumar Singh*

Alagappa University, Gene Center, India
*Corresponding Author

38

Mirko Minini

The activation of miR-125a-5p/IP6K1 axis in breast cancer cells upon treatment with myo-Inositol

Mirko Minini[1,2], Sara Proietti[1,2], Noemi Monti[1,2], Alice Senni[1,2], Andrea Fuso[1], Alessandra Cucina[2], and Mariano Bizzarri[1]

[1)] Department of Experimental Medicine, Systems Biology Group, Sapienza University of Rome, viale Regina Elena 324, 00161 Rome, Italy; [2)] Department of Surgery "Pietro Valdoni", Sapienza University of Rome, Via Antonio Scarpa 14, 00161 Rome, Italy;

39

Aleksandra Jarmolowicz

Finding New Influenza Inhibitors by Small Molecules Docking to Genomic Influenza RNA Structural Motif

Aleksandra Jarmolowicz[1], Joanna Miszkiewicz[2], Joanna Trylska[2], Elzbieta Kierzek[1*]

[1)] Institute of Bioorganic Chemistry Polish Academy of Sciences, Poznan, Poland;
[2)] Centre of New Technologies, Warsaw, Poland
e-mail: *elzbieta.kierzek@ibch.poznan.pl*

40

Pietro Busko

Analysis of the Interaction between RNA-Binding Protein SAM68 and Scaffold Protein ITSN1 *in vitro*

BUSKO P., PANKIVSKYI S.V., SENCHENKO N.V., RYNDITCH A.V.

Institute of molecular biology and genetics, National Academy of Science of Ukraine, Kyiv
e-mail: *petrobusko80@gmail.com*

41

Martin Cagala

L1210 mouse leukemic cell line resistance to endoplasmic reticulum stress inducers.

Martin Cagala[1], Lucia Pavlíková[1], Mário Šereš[1], Žaneta Boleková[2], Karolína Kadlečíková[2], Albert Breier[1,2], and Zdena Sulová[1]

[1)] Department of Biochemistry and Cytochemistry Institute of Molecular Physiology and Genetics, Centre of Biosciences Slovak Academy of Sciences, Dúbravská cesta 9, 845 05 Bratislava; [2)] Institute of Biochemistry and Microbiology, Faculty of Chemical and Food Technology, Slovak University of Technology, Radlinského 9, 812 37 Bratislava

---

42

Konstantin Chernov

Introducing inducible fluorescent split cholesterol oxidase to mammalian cells.

Chernov KG[1], Neuvonen M[2,3], Brock I[2,3], Ikonen E[3,4], Verkhusha VV[5,6]

[1)] Department of Biochemistry and Developmental Biology; [2)] Department of Anatomy, Faculty of Medicine, University of Helsinki; [3)] Minerva Foundation Institute for Medical Research, Helsinki and [4)] Department of Anatomy, Faculty of Medicine, University of Helsinki, Helsinki 00290, Finland, *elina.ikonen@helsinki.fi*;
[5)] Department of Biochemistry and Developmental Biology and vladislav.verkhusha@einstein.yu.edu;
[6)] Department of Anatomy and Structural Biology, Albert Einstein College of Medicine, Bronx, New York 10461

43

Elijah Kolawole Oladipo

Comparative Structural Analysis of 3-D Predicted Matrix Protein of Influenza A /Swine/H1N2/Ibadan/2014 and A/Swine/H5N1/Ogbomoso/ 2014 in Nigeria

Oladipo E.K.[1,2*] and Oloke J.K.[2]

[1)] Department of Microbiology, Virus Research Laboratory, Adeleke University, Ede, Osun State, Nigeria; [2)] Department of Pure and Applied Biology (Microbiology / Virology Unit), Ladoke Akintola University of Technology, Ogbomoso, Oyo State, Nigeria.

44

Dian Jamel Salih

Cytogenetic effects of Anabolic steroid hormones on bodybuilder athlete in Duhok province - Kurdistan region of Iraq

Dian Jamel Salih[*1], Asaad Abdulwahid Alasady[2]

[1,2)] Department of Anatomy, College of Medicine/ University of Duhok
* Corresponding author: *dian.jamel@uod.ac*

45

Elinor Breiner-Goldstein

Semisynthetic macrolides can overcome resistance mechanisms.

Elinor Breiner-Goldstein, Zohar Eyal, Donna Matzov, Yehuda Halfon, Ella Zimmerman, Haim Rozenberg, Anat Bashan and Ada Yonath

The Department of Structural Biology, Weizmann Institute of Science, Rehovot, Israel

---

46

Hugo Garro

Inhibitory interaction of a Porphyrin derivative with the protein α-synuclein

GARRO H.A.[1], GENTILE I.[1], RAMUNNO C.F.[1], GONZÁLEZ N.[1], DELGADO-OCAÑA S.[1], GRIESINGER C.[2], AND FERNÁNDEZ C.O[1,2]

[1)] Max Planck Laboratory for Structural Biology, Chemistry and Molecular Biophysics of Rosario (MPLbioR, UNR-MPIbpC) and Instituto de Investigaciones para el Descubrimiento de Fármacos de Rosario (IIDEFAR, UNR-CONICET), Universidad Nacional de Rosario, Ocampo y Esmeralda, S2002LRK Rosario, Argentina;
[2)] Department of NMR-based Structural Biology, MPIb-BPC, Am Fassberg 11, D-37077 Göttingen, Germany.

47

Amar Ranjan

Diagnostic dilemma in plasmablastic myeloma and plasmablastic lymphoma

Amar Ranjan, Harshita Dubey, Pranay Tanwar

48

Cabadak Hülya

The effect of combination of cholinergic agonist and antagonist on apoptosis in megacaryocytic differentiated K562 leukemia cells.

Cabadak Hülya, Aydın Banu

Department of Biophysics, Marmara University, Medical School, Istanbul, Turkey

---

49

Hekmat El Magdoub

Betanin ameliorates cisplatin-induced kidney injury through TGF-β/Wnt/Notch3 pathways: Involvement of miRNA-135a & -135b

Rania M Rahmo[1], Lobna H Khedr[1], Hekmat M. El Magdoub[2], and Mona F. Schaalan[3]

[1] Pharmacology & Toxicology Department; [2] Biochemistry Department;
[3] Pharmacy Practice and Clinical Pharmacy Department, Faculty of Pharmacy;
Misr International University, Cairo, Egypt

50

Esther Jimah

Investigation of Hepatitis C Virus among Blood Donors by Molecular Assay in Kwara State, Nigeria.

Jimah E.M[1,2*], Onariosakpobare F.O[1], Ashaka O.S[1], Omoare A.A.[1], Oladipo E.K[2,3], and Agbede O.O.[1]

[1] Department of Medical Microbiology and Parasitology, University of Ilorin, P.M.B 1515, Kwara State, Nigeria; [2] Department of Pure and Applied Biology, Ladoke Akintola University of Technology, P.M.B 4000, Ogbomoso, Oyo-state, Nigeria; [3] Department of Microbiology, Laboratory of Molecular Biology and Bioinformatics, Adeleke University, P.M.B 250, Ede, Osun State, Nigeria; * *esmojem@yahoo.co.uk*

51

Shuan-Su Chen and Wei-Hsuan Yu

Matrilysin Mediates Nucleolus Trafficking for γ-Catenin and p53 and Bulky-Ball-Like Nucleolus Assembly in Cancer Stem Cells

Shuan-su C. Yu[1] , Po-Tsang Huang[1], Wen-Hung Kuo[2], and Wei-Hsuan Yu[1*]

[1] Institute of Biochemistry and Molecular Biology, College of Medicine, National Taiwan University, Taipei 10051, Taiwan. [2] Department of Surgery, National Taiwan University Hospital. (*corresponding author: *whyu2004@ntu.edu.tw*)

52

Joanna Miszkiewicz

Designed RNA enzyme directed against bacterial mRNA

Joanna Miszkiewicz[1,2], Ksenia Maximova[2], Maciej Łukaszewicz[3], Edward Darżynkiewicz[2,3], and Joanna Trylska[2]

[1)]Inter-Faculty Individual Studies in Mathematics and Natural Sciences, University of Warsaw, Banacha 2C, 02-097 Warsaw, Poland; [2)] Centre of New Technologies, University of Warsaw, Banacha 2C, 02-097 Warsaw, Poland; [3)] Division of Biophysics, Institute of Experimental Physics, Faculty of Physics, University of Warsaw, Żwirki i Wigury 93, 02-089, Warsaw, Poland

---

53

Ana Carolina Schwarzer

Expression and Purification of Recombinant Human Wnt Signaling Modulators for Functional Analysis

Schwarzer, A.C.A.P.[1], Santos, D.C.M.[1], Zanchin, N.I.T.[1], and Carneiro, F.R.G.[1,2]

[1)] Laboratory of Structural Biology and Protein Engineering, Instituto Carlos Chagas, FIOCRUZ, Paraná (Brazil);
[2)] Center For Technological Development in Health (CDTS), FIOCRUZ, Rio de Janeiro (Brazil)

54

Indrani Mukherjee

Lipid dependent oligomerization of the ENTH domain drives membrane remodeling during clathrin mediated endocytosis

Benjamin Kroppen[1], Indrani Mukherjee[1], Nelli Teske[2], Garima Jaipuria[3], Markus Zweckstetter[3], Claudia Steinem[2], and Michael Meinecke[1]

[1)] Department of Cellular Biochemistry, University Medical Center Göttingen, Humboldtallee 23, 37073 Göttingen; [2)] Institute of Organic and Biomolecular Chemistry, University of Göttingen, Tammannstrasse 2, 37077 Göttingen; [3)] MPI-BPC, Department of NMR-based Structural Biology, Göttingen

---

55

Sofiia Reshetniak

The dynamic organization of proteins in the synaptic bouton

Sofiia Reshetniak[1,2], Jan-Eike Ußling[1], Burkhard Rammner[1], Thomas Schikorski[3], Sven Truckenbrodt[1,2], and Silvio O. Rizzoli[1]

[1)] University Medical Center Göttingen, Institute for Neuro- and Sensory Physiology and Biostructural Imaging of Neurodegeneration (BIN) Center, Göttingen, 37073, Germany;
[2)] International Max Planck Research School for Molecular Biology, Göttingen, Germany;
[3)] Department of Neuroscience, Universidad Central del Caribe, Bayamon, Puerto Rico, United States of America.

56

James Allen de Borja

Orthogonal assays and high-content imaging eliminate artefactual bioactivities in cardiometabolic natural products early drug screens

James Allen de Borja, Marian Abigaile N. Manongdo, Carmela Rieline Cruz, Aileen Geobee Uy, Shela Marie Algodon, Arman Ghodsinia, Charles Christopher Bataclan, Ryan Timothy Yu, and Reynaldo L. Garcia

Disease Molecular Biology and Epigenetics Laboratory, National Institute of Molecular Biology and Biotechnology, University of the Philippines Diliman, Quezon City 1101, Philippines

---

57

Sidney Allen Chua

Investigating the role of N-TAF1 and aberrant splicing in the pathogenesis of X-linked Dystonia Parkinsonism

Antoni Andreu Martija, Sidney Allen Chua, Jana Quismundo, Jose Gabriel Hilario, Jose Lorenzo Ferrer, Marie Isabelle Viola, and Reynaldo Garcia

Disease Molecular Biology and Epigenetics Laboratory, National Institute of Molecular Biology and Biotechnology, University of the Philippines Diliman

---

58

Evgeniia Ulas

The effect of nuclear receptors LXR and RXR agonists treatment on small GTPase ARL7 level in cultured HeLa and Vero cells

E.W. Ulas[1], E.S. Nadezhdina[1], and A.V. Burakov[2]

[1)] Institute of Protein Research of RAS, Pushchino, Russia;
[2)] A.N. Belozersky Research Institute of Physico-Chemical Biology MSU, Moscow, Russia

---

59

Yelyzaveta Zhyr

Changes in the Quantitative Composition of Plant Glycolipids during the Pathogenesis

Yelyzaveta Zhyr, Viktoriia Belava, and Olha Panyuta

Taras Shevchenko National University of Kyiv, ESC Institute of Biology and Medicine, Ukraine

---

60

Vasyl Velykyi

Antitumor Effectiveness of the Liposomal Form of the Cytotoxic Bacterial Lecitin of Bacillus Subtilis B-7025

V. Velykyi, O. Dvorshchenko, O. Garashchenko, O. Kruts, and G. Didenko

Taras Shevchenko National University, Kiev, Ukraine

61

N. Frenkel

Chromatin modification effects on DNA replication

N. Frenkel, N. Barkai

Department of Molecular Genetics, Weizmann Institute of Science

---

62

Mohamed A. Hassan

Production and Characterization of Metallo-Serine Keratinase from *Bacillus amyloliquefaciens* with Prospective Application in Leather and textile Industries

Mohamed A. Hassan

Protein Research Department, Genetic Engineering and Biotechnology Research Institute (GEBRI), City of Scientific Research and Technological Applications (SRTA-City), New Borg El-Arab City, P.O. Box: 21934 Alexandria, Egypt. E-mail: *madel@srtacity.sci.eg*

63

Volodymyr Kavetskyi

Cloning and structural analysis of abiotic stress tolerance gene in *Deschampsia antarctica* E. Desv.

Kavetskyi V., Mel'nyk V. M., Andreev I. O., Myryuta G. Yu., and Kunakh V.A.

Institute of Molecular Biology and Genetics of NASU, 150, Zabolotnogo Str., Kyiv, Ukraine, 03143; Educational and Scientific Center "Institute of Biology and Medicine" of Taras Shevchenko National University of Kyiv, Kyiv, Hlushkova Avenue, 2, Kyiv, Ukraine, 03127

64

Steve Thorne

The Copernican Project

Steve Thorne

www.celll.org, Berkeley, California, USA

65

Lena A. K. Bögeholz

Slow deformylation of nascent proteins on the ribosome

Lena A. K. Bögeholz, Evan Mercier, and Wolfgang Wintermeyer

Research Group *Ribosome Dynamics*, MPI for Biophysical Chemistry, Göttingen

---

66

Sergey Zaitsev

Regulation of the Lipase Activity in Complexes with the Synthetic and Natural Polymers

A.A. Savina[1], L.S. Garnashevich[1], I.S. Zaitsev[1], D.S. Muzychenko[1], M.A. Ovcharova[1], M.S. Tsarkova[1], and S.Yu. Zaitsev[1,2]

[1)] Moscow SAVMB, Acad. Scriabin Str. 23, Moscow109472, Russia;
[2)] Institute of Bioorganic Chemistry RAS, Miklukho-Maklaya Str. 16/10, Moscow177997, Russia. E-mail: *s.y.zaitsev@mail.ru*

67

Ismail El Shaffei

Ameliorative Effect of Betanin on Experimental Cisplatin-Induced Liver Injury; a Focus on SIRT-1/PGC-1α Signaling Pathways

Ismail El Shaffei, Rania Salama, Sahar Raouf, Doaa Boshra and Mona Schaalan

Faculty of Pharmacy, Misr International University (MIU), Cairo, Egypt

---

68

Katarzyna Więciorek

Defective endocytosis leads to mitochondrial malfunction

Katarzyna Więciorek[1,2,3], King Faisal Yambire[1,3], Ángela Sánchez Guerrero[1,3], Ira Milosevic[3], and Nuno Raimundo[1]

[1)] Institute of Cellular Biochemistry, University Medical Center, Göttingen, Germany;
[2)] Doctoral program of Molecular Medicine, University of Göttingen, Germany;
[3)] European Neuroscience Institute Göttingen – A Joint Initiative of the University Medical Center Göttingen and the Max-Planck-Society, Göttingen, Germany
k.wieciorek@stud.uni-goettingen.de

69

Simon Henriet

Intron creation by miniature transposons and the emergence of a novel splicing code

Simon Henriet, Berta Colom Sanmartí, Sara Sumic, and Daniel Chourrout

University of Bergen, Sars International Centre for Marine Molecular Biology, Bergen, Norway

---

70

Prashant Rawat

Role of Stress Induce Phase Separation in Transcriptional Attenuation

Prashant Rawat[1], Marc Böhning[2], Fernando Aprile Garcia[1], Barbara Hummel[1], Nathalie Eisenhardt[1], Gerhard Mittler[1], Jorma J. Palmivo[3], Andrea Pichler[1], Patrick Cramer[2], and Ritwick Sawarkar[1]

[1)] Max Planck Institute of Immunobiology and Epigenetics, Freiburg, Germany,
[2)] Max Planck Institute for Biophysical Chemistry, Göttingen, Germany;
[3)] Institute of Biomedicine, University of Eastern Finland, Kuopio, Finland

---

71

Harshita Dubey

Clinical Evaluation of Procalcitonin in Gastrointestinal Cancer

---

72

Kristina Simonicova

Transport and Metabolism of Hypomethylating Agents and Development of Resistance

Kristina Simonicova[1], Lubos Janotka[1], Helena Kavcova[1], Zdena Sulova[1], Albert Breier[1,2], Lucia Messingerova[1,2]

[1)] Department of Biochemistry and Cytochemistry, Institute of Molecular Physiology and Genetics, Centre of Biosciences, Slovak Academy of Sciences, Bratislava, Slovak Republic;
[2)] Institute of Biochemistry and Microbiology, Faculty of Chemical and Food Technology, Slovak University of Technology, Bratislava, Slovak Republic

---

73

Khubaid Ur Rehman

TBA

Khubaid Ur Rehman

National Control Laboratory for Biologicals, Drug Regulatory Authority of Pakistan, Islamabad, Email: *khubaid@gmail.com*

74

Aleksandr Kurahenko

New Tool for the TADs Prediction

Aleksandr Kurahenko

Moscow State University, Russia

75

Natia Tepnadze

Study of genetic diversity of *Aegilops tauschii* Coss. samples spread in Georgia using the chloroplast DNA

Natia Tepnadze[1], Mari Gogniashvili[2]

[1)] Agricultural University of Georgia,[2)] Agricultural University of Georgia

---

76

Eugenia Solomatina

Focal adhesion dynamics: ROCK and MLCK have a similar effect in normal and tumor cells

Solomatina E. S.*, Kovaleva A. V., and Saidova A. A.

Biology Department of Lomonosov Moscow State University
*Email: *solomatinaj@gmail.com*

77

Angel J. Santiago-Lopez

Mobilization of cellular stress signaling in protein misfolding diseases for controlled gene delivery

Angel J. Santiago-Lopez[1,2,3], Alex Nazzari[3], Ken Berglund[3], Claire-Anne Gutekunst[3], and Robert E. Gross[1,3]

[1)] Interdisciplinary Bioengineering Program, Georgia Institute of Technology, Atlanta, GA
[2)] School of Chemical and Biomolecular Engineering, Georgia Institute of Technology, Atlanta, GA; [3)] School of Medicine Department of Neurosurgery, Emory University, Atlanta, GA

78

Elena Ryabova

Swiss Cheese Gene Knockdown Leads to a Glial Cells death in the Adult *Drosophila* Brain

Ryabova E., Surina N., Zhmuydina D., Melentev P., and Sarantseva S.V.

B.P. Konstantinov Petersburg Nuclear Physics Institute, National Research Centre «Kurchatov Institute», Gatchina, Russia

---

79

Nina Surina

Amyloid Beta Peptide Influence on Morphology and Functions of *Drosophila Melanogaster* Glial Cells

Surina N.V.[1], Ryabova E.V.[1], Zhmuydina D.R.[1], and Sarantseva S.V.[1]

Petersburg Nuclear Physics Institute named by B.P. Konstantinov of National Research Centre «Kurchatov Institute», Gatchina, Russia, *anilannas123@gmail.com*

80

Mizanurfakhri Ghazali

Comparative expression of the Microfold (M) cells between the alginate hydrogel beads and transmembrane 3D culture system.

Mizanurfakhri Ghazali[1,2*], Sharaniza Ab-Rahim[2], and Mudiana Muhamad[2]

[1)] Institute of Medical and Molecular Biotechnology (IMMB), Faculty of Medicine, Universiti Teknologi MARA, Cawangan Selangor, Kampus Sungai Buloh, 47000 Sungai Buloh, Selangor, Malaysia; [2)] Department of Biochemistry and Molecular Medicine, Faculty of Medicine, Universiti Teknologi MARA, Cawangan Selangor, Kampus Sungai Buloh, 47000 Sungai Buloh, Selangor, Malaysia; *Corresponding author Tel.: +60361267360; email: *mizanurfakhrighazali@yahoo.com*

81

Zuzana Kočibálová

Atypical Expression of Latrophilin-1 in Acute Myeloid Leukemia: Balance between Multidrug Resistance and Immune Escape?

Zuzana Kočibálová[1], Martina Guzyová[1], Miroslava Sojková[1], Denisa Imrichová[2], Zdena Sulová[2], and Albert Breier[1,2]

[1)] Institute of Biochemistry and Microbiology, Faculty of Chemical and Food Technology, Slovak University of Technology in Bratislava, Slovak Republic; [2)] Department of Biochemistry and Cytochemistry, Institute of Molecular Physiology and Genetics, Centre of Biosciences SAS, Slovak Academy of Sciences, Bratislava, Slovak Republic

---

82

Martin Škandík

Quercetin-Quinone Derivative Mitigates Microglial Inflammation and Senescence Markers by Redox Homeostasis Modulation

Martin Škandík, Lucia Račková

Centre of Experimental Medicine, Institute of Experimental Pharmacology & Toxicology, Slovak Academy of Sciences, Bratislava, Slovakia

---

83

Marija Lj Medar

Immobilisation Stress Changes Expression Pattern of Core Clock and Steroidogenic Genes in Leydig Cells

Marija Lj Medar, Aleksandar Z Baburski, Silvana A Andric, and Tatjana S Kostic

University of Novi Sad, Faculty of Sciences Novi Sad, Department of Biology and Ecology, Laboratory for Reproductive Endocrinology and Signaling Group, Novi Sad, Serbia

---

84

Isidora Starovlah

Signaling Pathways Regulating the Mitochondrial Dynamic and Acrosomal Reaction are Disturbed in Spermatozoa from Stressed Adult Rats

Starovlah IM, Radovic SM, Kostic TS, and Andric SA

Laboratory for Reproductive Endocrinology and Signaling, Laboratory for Chronobiology and Aging, Centre of Excellence CeRES, Faculty of Sciences, University of Novi Sad, Novi Sad, Serbia

---

85

Jeremy McCallum-Loudeac

Sex Driven Changes in the Spinal Cord over Puberty and Adolescence: The Role of Hormones on Gene Expression

Jeremy McCallum-Loudeac[1], Greg M. Anderson[1,2], and Megan J. Wilson[1]

[1)] Department of Anatomy, University of Otago, Dunedin, New Zealand;
[2)] Centre for Neuroendocrinology, University of Otago, Dunedin, New Zealand

---

86

Malgorzata Majewska

Characterisation of *HMGR4* promoter from *Salvia miltiorrhiza*

Malgorzata Majewska[1], Renata Grabkowska[1], Agnieszka Jelen[3], Ewa Balcerczak[3], Lukasz Kuzma[1], and Piotr Szymczyk[2]

[1)] Department of Biology and Pharmaceutical Botany, Medical University of Lodz, Muszynskiego 1, 90-151 Lodz, Poland, *malgorzata.majewska1@stud.umed.lodz.pl*.
[2)] Department of Pharmaceutical Biotechnology, Medical University of Lodz, Muszynskiego 1, 90-151 Lodz, Poland; [3)] Department of Pharmaceutical Biochemistry and Molecular Diagnostics, Medical University of Lodz, Muszynskiego 1, 90-151 Lodz, Poland

---

87

Preethi Basavaraju

TBA

Preethi Basavaraju

Bharathiar University, Coimbatore, India

88

Michał Marcinkowski

New partners of FTO protein – biophysical and biochemical studies

Michał Marcinkowski[1], Tomasz Pilżys[1], Damian Garbicz[1], Jan Piwowarski[1], Maria Winiewska[1], Kaja Przygońska[1], Maciej Gielnik[2], Elżbieta Grzesiuk[1], and Jarosław Poznański[1]

[1)] Institute of Biochemistry and Biophysics, PAS, Pawińskiego 5a, 02-106, Warsaw, Poland,
[2)] Department of Macromolecular Physics, Faculty of Physics, Adam Mickiewicz University, Umultowska 85, Poznań, Poland

---

89

Alina Stein

Protein Engineering of Artificial Metalloenzymes with Unnatural Amino Acids and Rational Design

Alina Stein[1], Marc Masmünster[1], Alexandria Deliz Liang[1], Johannes G. Rebelein[1], Jaume Bonet[2], Bruno Correia[2], and Thomas R. Ward[1]

[1)] University of Basel, Department of Chemistry;
[2)] École Polytechnique Fédérale de Lausanne, Institute of Bioengineering

---

90

Nino Rostiashvili

Thrombocytosis in JAK2 V617F positive myeloproliferative neoplasms

Nino Rostiashvili, Giorgi Kharebava

Allele Ltd, Tbilisi, Georgia
Corresponding author: Giorgi Kharebava, PhD, Allele Ltd, Lisi Lake #4, Tbilisi, 0186, Georgia

---

91

Pravin Dudhagara

Microbiome study of sediment and water of polluted site by column based enrichment culture designed for the pesticides degradation

Pravin Dudhagara* and Rajesh Patel

Bioinformatics and Supercomputer Lab., Department of Biosciences (UGC-SAP-DRS-II & DST-FIST-I), Veer Narmad South Gujarat University, Surat
*dudhagarapr@gmail.com

92

Julie Trolle

Engineering Prototrophy In Mammalian Cells

Julie Trolle

New York University, USA

93

Nina V. Voronova

Five new mitochondrial genomes of aphids: Aphis affinis, A. craccivo-ra, A. fabae mordvilkoi, Appendiseta robiniae and Therioaphis tenera

Nina V. Voronova

Belarusian State University, Minsk, Belarus

94

Zehra Kanlı

Investigation of the Role of M3 muscarinic receptor in Chronic Myeloid Leukemic Cells

Zehra Kanlı, Banu Aydın, and Hülya Cabadak

Department of Biophysics, Marmara University, Medical School, Istanbul, Turkey

95

Nataliia Chornenka

Investigation of excess collagen formation in the conditions of alkaline esophageal burn and using of melanin

Chornenka N.M., Raetska Ya. B., and Savchuk O.M

Taras Shevchenko National University of Kyiv, Educational-scientific centre " Institute of Biology and Medicine", Department of Experiment Biology, 64 Volodymirska St., 01033 Kyiv, Ukraine; e-mail: *nata.chornenka24@gmail.com*

96

Tomas Lagunas

A massively parallel reporter assay to investigate the contribution of noncoding variation in autism spectrum disorder

Tomas Lagunas Jr.[1,2,3], Stephen Plassmeyer[1,2], Joseph D. Dougherty[1,2]

Departments of Genetics[1] and Psychiatry[2], Molecular Genetics & Genomics PhD program in the Division of Biology and Biomedical Sciences[3] at Washington University School of Medicine, St. Louis, MO 63110

97

Aster E. Pijning

TBA

Aster E. Pijning, Joyce Chiu, Emma Ramsay, and Philip J. Hogg
University of Sydney, New South Wales, Australia

98

Sundararaghavan Pattabiraman

Vimentin protects differentiating stem cells from stress

Sundararaghavan Pattabiraman, Gajendra Kumar Azad, Triana Amen, Eran Meshorer, and Daniel Kaganovich

University of Göttingen, Center for Nanoscale Microscopy and Molecular Physiology of the Brain, Department of Experimental Neurodegeneration

99

Ruchika

TBA

Ruchika

Japan Advanced Institute of Science and Technology, Japan

100

Temitope C. Ekundayo

Molecular characterisation and differential numerical delineation of DNA fingerprints of Plesiomonas shigelloides isolates into strain clusters

Temitope C. Ekundayo*; Anthony I. Okoh

SAMRC Microbial Water Quality Monitoring Centre, University of Fort Hare, Alice, Eastern Cape, South Africa;
cyruscyrusthem@gmail.com / +27604890881

101

Swarna Kanchan

Responses of antioxidants to mitigate oxidative stress in cyanobacterium Scytonema spp.

Swarna Kanchan[1], Rajeshwar P. Sinha[2], and Minu Kesheri[3*]

[1)] INRIA, Bordeaux sud-ouest, Talence-33400, France; [2)] Laboratory of Photobiology and Molecular Microbiology, Centre of Advanced Study in Botany, Banaras Hindu University, Varanasi-221005 India; [3)] Amity University, Ranchi-834002, India

---

102

Minu Kesheri

Temporal variations of Superoxide dismutase in bloom forming cyanobacterium Microcystis aeruginosa

Minu Kesheri[1], Rajeshwar P. Sinha[2], and Swarna Kanchan[3*]

[1)] Amity University, Ranchi-834002, India; [2)] Laboratory of Photobiology and Molecular Microbiology, Centre of Advanced Study in Botany, Banaras Hindu University, Varanasi-221005 India; [3)] INRIA, Bordeaux sud-ouest, Talence-33400, France.
*Corresponding author: Tel.: +33-605674989; e-mail: *swarnabioinfo@gmail.com*

103

Niklas Mueller

Chemogenetic control of acute H2O2 production in cells with D-amino acid oxidase (DAAO)

Niklas Mueller, Kurt Noack, Pedro F. Malacarne, Katrin Schröder, Flávia Rezende, and Ralf P. Brandes

Institute for Cardiovascular Physiology, Goethe-University, Frankfurt, Germany; German Center of Cardiovascular Research (DZHK), Partner Site Rhein Main, Frankfurt, Germany

---

104

Parastoo Karimian

Functional analysis of the Arabidopsis non-characterized Leucine-rich receptor-like kinase, LRR2, interacting with defence-related G-proteins

Parastoo Karimian, Tam Thi Thanh Dang, Yuri Trusov, and Jose Botella

The University of Queensland, Australia

105

Gabriel Vusanimuzi Nkomo

TBA

Gabriel Vusanimuzi Nkomo

Central University of Technology Free State, South Africa

106

Maria Fando

SmAP from Halobacterium salinarum has two different forms with the same structure but a different RNA-binding affinity

M. S. Fando[1], N. V. Lekontseva[1], J. A. Buyuklyan[1,2], and A. D. Nikulin[1]

[1)] Institute of Protein Research, Russian Academy of Sciences, Pushchino, Russia;
[2)] South Ural State Humanitarian Pedagogical University

107

Triana Amen

Stress Granules regulate cell metabolism through clustering of mitochondrial membrane protein VDAC

Triana Amen, Daniel Kaganovich

Universitätsmedizin Göttingen, Germany
